AERATION
A WASTEWATER
TREATMENT PROCESS

Water Pollution Control Federation
American Society of Civil Engineers

WPCF—Manual of Practice—No. FD-13
ASCE—Manuals and Reports on Engineering Practice—No. 68

Prepared by a
Joint Task Force of the Water Pollution Control Federation and the American Society of Civil Engineers
The Joint Task Force was composed of members of the ASCE Task Committee on Aeration of the Environmental Engineering Division Committee on Water Pollution Management and the WPCF Facilities Development Subcommittee of the Technical Practice Committee

Published by the
Water Pollution Control Federation
601 Wythe Street
Alexandria, Virginia 22314-1994

and the

American Society of Civil Engineers
345 East 47th Street
New York, New York 10017-2398

The material presented in this publication has been prepared in accordance with generally recognized engineering principles and practices, and is for general information only. This information should not be used without first securing competent advice with respect to its suitability for any general or specific application.

The contents of this publication are not intended to be and should not be construed to be a standard of the American Society of Civil Engineers (ASCE) or the Water Pollution Control Federation (WPCF) and are not intended for use as a reference in purchase specifications, contracts, regulations, statutes, or any other legal document.

No reference made in this publication to any specific method, product, process, or service constitutes or implies an endorsement, recommendation, or warranty thereof of ASCE or WPCF.

ASCE and WPCF make no representation or warranty of any kind, whether expressed or implied, concerning the accuracy, completeness, suitability, or utility of any information, apparatus, product, or process discussed in this publication, and assume no liability therefore.

Anyone using this information assumes all liability arising from such use, including but not limited to infringement of any patent or patents.

Library of Congress Cataloging in Publication Data
Aeration in wasterwater treatment.

(ASCE manuals and reports on engineering
practice; no. 68) (WPCF manual of practice; no. FD-13.)
   Bibliography: p.
   Includes index.
   1. Sewage—Purification—Aeration.   I. American
Society of Civil Engineers.   II. Water Pollution
Control Federation.   III. Series.   IV. Series:
WPCF manual of practice; FD-13.
TD201.W337 no. FD-13   [TD758]   628.1'68 s   88-20856
                                  [628.3'51]

   includes bibliographical references and index.

   1. Aeration—Wastewater Treatment.   I. American Society of
Civil Engineers.   II. Water Pollution Control Federation.
III. WPCF manual of practice; no. FD-13.

Joint Task Force on Aeration in Wastewater Treatment

William C. Boyle, Chairman
Richard Atoulikian
Charles R. Baillod
Edwin L. Barnhart
William L. Berk
Arthur G. Boon
Richard C. Brenner
Haskal Brociner
Linfield C. Brown
Paul Busch
Hugh J. Campbell, Jr.
Paul T. Carver
Glen T. Daigger
David Di Gregorio
W. Wesley Eckenfelder, Jr.
Lawrence A. Ernest
Lloyd Ewing
R. Gary Gilbert
Mervin C. Goronszy
Michael G. Hardie
Gregory Huibregtse
Richard P. Johansen
Terry L. Johnson
Rolf Kayser
Boris Khudenko
Gerald A. Knazek
Paul Kubera
Richard Lehmann

M.B. Mandt
Frederick K. Marotte
James J. Marx
James J. McKeown
James A. Mueller
Wayne Paulson
George B. Powell
David C. Rasmussen
David T. Redmon
R. Bruce Ringrose
John Roeber
J. W. Gerald Rupke
F. Michael Saunders
Norbert W. Schmidtke
Walter Schuk
Gerald L. Shell
George W. Smith
Gordon Speirs
Vernon T. Stack, Jr.
Michael K. Stenstrom
Richard W. Stone
Richard B. Weber
Bruce R. Willey
Thomas E. Wilson
Jerome D. Wren
Shang Wen Yuan
Fred W. Yunt

Manuals of Practice for Water Pollution Control

(as developed by the Water Pollution Control Federation)

The WPCF Technical Practice Committee (formerly the committee on Sewage and Industrial Wastes Associations) was created by the Federation Board of Control on October 11, 1941. The primary function of the committee is to originate and produce, through appropriate subcommittees, special publications involving technical aspects of the broad interests of WPCF. These manuals are intended to provide background information through a review of technical practices and detailed procedures that research and experience have shown to be functional and practical.

Water Pollution Control Federation
Technical Practice Committee
Control Group

C.S. Zickefoose, *Chairman*
P.T. Karney, *Vice Chairman*
L.J. Glueckstein
A.J. Callier
F.D. Munsey
T. Popowchak

Authorized for publication by the Board of Control
Water Pollution Control Federation, 1987
Quincalee Brown, *Executive Director*

ABSTRACT

AERATION
A WASTEWATER TREATMENT PROCESS

This manual summarizes current aeration practices in wastewater treatment, and includes both theoretical and practical guidelines for the design and operation of such unit processes.

The first chapter describes the history of aeration and is followed by chapters on oxygen requirements, modeling, diffused air and mechanical aeration systems, process control, O & M, and aeration-testing protocols.

The manual is intended to be of practical use to the design engineer and is based on the experience of engineers in the field of wastewater treatment plant design and operation. Charts and illustrations are used throughout to reinforce the text.

Manuals and Reports on Engineering Practice

(as developed by the ASCE Technical Procedures Committee, July, 1930, and revised March, 1935, February, 1962, April, 1982)

A manual or report in this series consists of an orderly presentation of facts on a particular subject, supplemented by an analysis of limitations and applications of these facts. It contains information useful to the average engineer in his everyday work, rather than the findings that may be useful only occasionally or rarely. This manual is not, however, in any sense a "standard;" nor is it so elementary or so conclusive as to provide a "rule of thumb" for nonengineers.

Furthermore, material in this series, in contrast with a paper (which expresses only one person's observations or opinions), is the work of a committee or group selected to assemble and express information on a specific topic. The committee is under the general direction of one or more of the technical divisions and councils, and the product is subjected to review by the Executive Committee of that division or council. As a step in this review, proposed manuscripts are often brought before the members of the technical divisions and councils for comments, which may serve as the basis for improvement. When published, each work shows the names of the committee by which it was compiled; also, indicated clearly are the several review processes through which it was passed so that its merit may be definitely understood.

In February, 1962 (revised in April, 1982), the Board of Direction voted to establish: A series titled "Manuals and Reports on Engineering Practice," including the manuals published and authorized to date, future Manuals of Professional Practice, and Reports on Engineering Practice. All such manual or report material of the Society would have been refereed in a manner approved by the Board Committee on Publications and would be bound, with applicable discussion, into books similar to past manuals. Numbering would be consecutive and would be a continuation of present manual numbers. In cases of joint committee reports, bypassing of Journal publications may be authorized.

American Society of Civil Engineers
Water Pollution Management Committee
Control Group

R.J. Nogaj, Chairman
S.C. Reed, Vice-Chairman
G.W. Bryant, Secretary
D.L. Glasgow
F.M. Saunders

Authorized for publication by the Board of Direction
American Society of Civil Engineers, 1988
E. Phrang, *Executive Director*

Preface

The Subcommittee on Air Diffusion was created by the WPCF Committee on Sewage and Industrial Wastes Practice (now the Technical Practice Committee) in 1942. This subcommittee produced Manual of Practice No. 5, "Air Diffusion in Sewage Works," in 1952.

As technical practices, new equipment availability, and process sophistication progressed, the scope and depth of the material covered in this manual was broadened. The subcommittee was reactivated in 1965; it published a revised version of the text in 1971.

In 1983, a joint committee of the WPCF and the ASCE was formed to revise the 1971 edition of the manual.

This manual should be considered by the practicing engineer as an aid and a checklist of items for wastewater aeration, as represented by acceptable current procedure. It is not intended to be a substitute for engineering experience and judgment, nor as a treatise replacing standard texts and reference material.

Similar to other manuals prepared on special phases of engineering, the manual recognizes that this field of engineering is constantly progressing with new ideas, materials, and methods coming into use. It is hoped that users will present any suggestions for improvement to the Technical Practice Committee of the WPCF or the Environmental Engineering Division of the ASCE for possible inclusion in future revisions.

The members of the joint committee thank the reviewers of the manual for their assistance in submitting their suggestions for improvement. In addition to members of the Task Force, the WPCF Technical Practice Committee Control Group, and the ASCE Committee on Water Pollution Management, reviewers included Albert B. Pincince.

Thanks also are due to Harry Tuvel, formerly with ASCE staff, for his contributions to the manual.

Technical assistance was provided by Eugene De Michele, with editorial assistance by Carole C. Hayward.

Contents

Chapter 1

Introduction

In the 15 years since the last publication of the Manual of Practice No. 5, *Aeration in Wastewater Treatment*, substantial change has taken place in oxygen transfer systems design in the U.S. The major impetus for this change was the energy crisis of the early 1970s; interest and awareness in the sanitary-engineering community relative to aeration system efficiency was rekindled. Wastewater aeration represents one of the most energy intensive operations in the wastewater treatment system, consuming between 50 and 90% of the total energy cost of a typical municipal treatment facility.[1] A general data survey made available in 1982 on municipal and industrial wastewater treatment installations shows that, on the North American continent, there are approximately 1.3 million kW of aeration equipment in place at an installed value of $0.6 to 0.8 billion.[2] Operating costs for these systems may be expected to be about $0.6 billion per yr in 1982 dollars.

The testing of oxygen transfer devices in both clean and process waters is a recent major change that has taken place in this field. Presently, a consensus standard for clean waters has been adopted by a large segment of the industry.[3] Extensive aeration equipment tests in process waters using new, more accurate techniques have led to a better understanding of the translation of clean water test results to process conditions.[4] A renewed interest in fine pore (fine bubble) aeration in the U.S. has instigated new equipment development and a multiplicity of new maintenance considerations. Field studies that have demonstrated the importance of diffuser placement and tank geometry have produced more efficient system designs. The development of numerous control strategies and the redesign of prime movers have also improved aeration efficiency.

*O*BJECTIVES

Summarizing aeration practices in wastewater treatment today is the major objective of this manual. The current status of selection, design, operation, maintenance testing, and control of oxygen transfer systems has been reviewed. The design practices cited are not standards but represent consensus practice used today. Because of the rapid changes in technology that are currently taking place, it is possible that some systems have been omitted from discussion; omission does not, however, imply disapproval. Chapters on estimating oxygen requirements, modeling, diffused and mechanical aeration systems, control, and O &

M are directed at improving facility design. To understand new concepts and applications, however, the reader will find that a general knowledge of aeration design history is useful.

Early developments in aeration

Experiments on wastewater aeration started in England as early as 1882[5]; they continued on aeration and aerated filter wastewater treatment in England and the U.S. until the activated sludge process was revealed to the world by Arden and Lockett[6] (who were under the direction of Gilbert) on April 3, 1914.

In the early experiments, air was introduced through open tubes or perforations. In 1904, a patent was granted to Henderson in Great Britain for a perforated metal plate diffuser. Other patents were granted, primarily in Great Britain, before and after this for diffuser devices of various types and materials, such as porous tubes with fibrous material in the annular space between the tubes, and various nozzles.[7]

As the activated sludge investigations progressed, greater air economy was sought with smaller bubbles produced by porous diffuser media. Porous tile was first used by Fowler in his laboratory experiments.[5,6] About 1914, experiments were made in England using natural limestone, fire brick, sand and glass mixtures, pumice, and other materials; most of the materials tried were too dense. In 1915, Jones and Attwood, Ltd., offered a porous plate set in a cast iron box. The plates were made of concrete by a secret process; they were commonly used in Great Britain and its colonies until the patents expired. Filter plates were used before 1910 in the cyanide and chemical processes. Their use for wastewater treatment was suggested by Bartow in 1915.

In following years, in the U.S., Carborundum Co., Ferro Corp., and Norton Co. offered porous plates.[7] Experiments with various porous media and other media types are still continuing.

Diffuser clogging was a problem in numerous cases from the earliest aeration experiments. At Salford in 1915, perforated pipes were abandoned in favor of "diffusers" because 90% of the jets had plugged.[7] With the advent of porous diffusers, clogging became a greater problem. Early work by Bushee and Zack[8] at the Sanitary District of Chicago from 1922 to 1924 prompted coarser media use to avoid severe clogging problems. Roe[9] outlined in detail numerous diffuser-clogging causes.

Although some diffuser clogging was apparently caused by liquor-side clogging, there were more instances in which the clogging stemmed from dirt particles in the air that caused clogging from the air side. This was especially true respecting diffuser media and other aeration devices in which emphasis was placed on small air bubble production. Much experimentation has been done with many filter types in an effort to produce air sufficiently clean to prevent this type of clogging. Today, internal, air-side clogging has virtually been eliminated by high-efficiency air filtration use.

Mechanical aeration was one answer to the clogging problem. In Sheffield, England, porous tile diffusers were clogged in a few months by excessive iron salt in the wastewater. This spurred the development of apparatus for aeration without porous diffuser use. In 1916, Archimedean screw-type aerators were used to build an experimental tank. In 1920, Sheffield used horizontal paddle wheels in narrow channels to build a large-scale plant. Down-draft and up-draft type mechanical aerators were developed in an effort to diffuse air efficiently and to solve the clogging problem. Bolton, working in conjunction with Ames Crosta Engineering Company, used a center cone to develop a surface up-flow aeration system. This unit was installed at Bury, England, in

1919.[10] Since that time, a multitude of mechanical-aerating devices have been developed and marketed. These devices serve an important function in many aeration applications today.

Another approach to the clogging problem was development of the large orifice-type diffuser. First developed in the early 1950s, these devices improved on earlier perforated pipes and were designed for easy maintenance and accessibility. In general, however, these devices generate larger bubbles, therefore, sacrificing substantial transfer efficiency for maintenance ease.

With the emphasis on more energy-efficient aeration systems in the 1970s, the U.S. turned to European technology. Fine pore air diffusion has subsequently gained renewed popularity as a highly competitive system in the current market. Yet, considerable concern has been registered regarding the performance and maintenance of fine pore diffusion systems because of their susceptibility to clogging.[4]

The aeration market is in a substantial state of flux in the U.S. today. Emphasis on high efficiency has led to many intensive research programs aimed at the evaluation of design, operation, and control processes relative to improving overall system performance.

AERATION APPLICATIONS IN WASTEWATER TREATMENT

Air or oxygen-enriched air is introduced into wastewater in several unit processes designed for contaminant removal. In some instances, the air serves as an inexpensive gas for mixing (aerated grit chamber); in others, air is used for oxygen that may be used either to satisfy the biochemical oxygen demand in biological treatment processes or to act as an agent in the oxidation of undesirable contaminants.

The major uses of aeration in wastewater treatment include preaeration, aerated grit removal, grease flotation, aerobic biological treatment (activated sludge, aerated lagoons, aerobic digestion, and a variety of aerobic fixed-film processes), and post aeration. These unit operations are covered in detail in several publications, most notably in the manuals of practice on "Wastewater Treatment Plant Design" and "Operation of Wastewater Treatment Plants."[11,12] The reader is encouraged to review these two publications for descriptions of the design and operation of the various unit processes within a wastewater facility.

REFERENCES

1. Wesner, G.M., et al., "Energy Conservation in Municipal Wastewater Treatment." EPA—430/9—77—011, U.S. EPA, Washington, D.C. (1977).
2. Barnhart, E.L., "An Overview of Oxygen Transfer Systems." In "Proceedings of Seminar Workshop in Aeration System Design, Testing, Operation and Control." EPA—600/9—85—005, U.S. EPA, Cincinnati, Ohio (1985).
3. American Society of Civil Engineers. "Measurement of Oxygen Transfer in Clean Water." ASCE Standard, Am. Soc. of Civ. Eng., New York (1984).
4. "Summary Report—Fine Pore (Fine Bubble) Aeration Systems." EPA—625/8—85—010, U.S. EPA, Cincinnati, Ohio (1985).
5. Martin, A.J., "The Activated Sludge Process," MacDonald and Evans, London, England (1927).

6. Arden, *et al.*, "Experiments in Oxidation of Sewage Without the Aid of Filters." *J. Soc. Chem. Ind.*, **33,** 10 (1914).
7. "Air Diffusion and Sewage Works." Manual of Practice No. 5, Water Pollut. Control Fed., Washington, D.C. (1952).
8. Bushee, R.J. and Zach, S.I., "Tests on Pressure Loss in Activated Sludge Plants." *Eng. News-Rec.*, **93,** 21 (1924).
9. Roe, F.C., "The Installation and Servicing of Air Diffuser Mediums." *Water and Sew. Wks.*, **81,** 115 (1934).
10. McKinney, R.E. and O'Brien, W.J., "Activated Sludge—Basic Design Concepts." *J. Water Pollut. Control Fed.*, **40,** 1831 (1968).
11. "Wastewater Treatment Plant Design." Manual of Practice No. 8, Water Pollut. Control Fed., Alexandria, Va.; Manual of Engineering Practice No. 36, Am. Soc. of Civ. Eng., New York, N.Y. (1977).
12. "Operation of Wastewater Treatment Plants." Manual of Practice No. 11, Water Pollut. Control Fed., Alexandria, Va. (1976).

Chapter 2

Oxygen Requirements in Biological Processes

Introduction

Oxygen requirements in biological systems are the result of three primary demands:

- Carbonaceous biochemical oxygen demand
- Nitrogenous biochemical oxygen demand
- Inorganic chemical oxygen demand

The design engineer must estimate not only the total oxygen demand caused by these sources, but also the spatial and temporal variations in the demands from the sources within the reactor system to be aerated. Finally, aeration systems also may serve to mix reactor contents. Although often not the most efficient way to maintain a well-mixed environment, the engineer needs to check whether the aeration system that is designed will also satisfy the mixing requirements of the process.

Calculation of Process Oxygen Requirements

Carbonaceous Biochemical Oxygen Demand. A number of approaches have been taken to estimate the oxygen requirements caused by the biochemical oxidation of organic matter. Today, many regulatory agencies specify oxygen design criteria for unit processes. Although the engineer generally is given latitude in process selection (providing documentation supports his or her design), agency requirements must be considered. Various empirical or rule-of-thumb techniques may be used to estimate the oxygen requirements for activated sludge systems. They include:

- 1500 cu ft to 2000 cu ft air/lb (94 to 125 m³/kg) of 5-day BOD applied[1]
- 1.1 lb oxygen transferred/lb (1.1 kg/kg) of peak 5-day BOD applied for conventional aeration tanks[1]
- 0.5 to 2.0 cu ft air/gal (3.7 to 15 m³/m³) of wastewater treated[2]
- 500 to 900 cu ft air/lb (31 to 56 m³/kg) of 5-day BOD removed, at least 3 cu ft air/min/ft (4.6 L/m · s) of tank for mixing[3]

Recently, more rational approaches based on the process oxygen balance have been developed. In the process oxygen balance, it is presumed that oxygen demand is the result of biochemical oxidation of the applied organics. This approach, typified by the Lawrence and McCarty[4] analysis of biological treatment systems, may be written for a steady-state treatment system:

Mass Oxygen Required

= Mass of Total Carbonaceous Oxygen Demand Utilized   (1)

− β Biomass VSS Wasted

where β represents the oxygen equivalent of cell mass, often calculated as 1.42 mass O_2/mass biomass VSS.[5] In Equation 1, it is important to note that organic matter is expressed as total carbonaceous oxygen demand (often described as ultimate BOD, theoretical oxygen demand, total biochemical oxygen demand, etc.). Therefore, it does not include all of the organics in the influent as measured by a COD test. There will be a significant fraction of particulate and soluble material that is not biodegradable. In this Chapter the biodegradable organic material will be designated by the letter S.

Rewriting Equation 1 symbolically,

$$R_c = Q(S_o - S) - \beta P_x \qquad (2)$$

Furthermore, based on biomass and organic substrate balances for a steady-state treatment system, it can be shown[4] that:

$$P_x = \frac{Q\,Y(S_o - S)}{1 + b\theta_c} \qquad (3)$$

Note, again, that P_x is the biomass produced, based on S, the total oxygen demand.

Therefore, substituting;

$$R_c = Q(S_o - S)\left(\frac{1 + b\theta_c - \beta Y}{1 + b\theta_c}\right) \qquad (4)$$

where

R_c = mass oxygen required per unit time to satisfy the carbonaceous biochemical oxygen demand, M/t

Q   = flow rate, L^3/t

S_o, S = total carbonaceous oxygen demand (ultimate BOD) of influent and effluent from aerated reactor, M/L^3

b   = endogenous decay coefficient based on biomass in aerated zone, 1/t

Y   = true cell yield (based on total carbonaceous oxygen demand), M/M

θ_c = solids retention time at steady state conditions, t

P_x = biomass production rate, M/t

Note that some engineers prefer to use observed yield, Y_{OB}, rather than the true yield, Y. They are related by Equation 5 as follows:

$$Y_{OB} = \frac{Y}{1 + \theta_c b} \qquad (5)$$

In Equation 4, b and Y are biological parameters representative of the biochemical system being aerated, θ_c is a process design parameter, and Q, S_o, and S are properties of the wastewater. Numerical values of Y depend on the units in which biomass and substrate are expressed. Typical values range from 0.25 to 0.40 g VSS/g COD removed.[6] Typical values for the decay coefficient, b, range from 0.04 to 0.075 per day.[4,6] Observed yield values are dependent on the solids retention time and tabulations are not readily available in the literature.

A slightly different approach has been taken by Eckenfelder[5] in describing the stoichiometry of carbonaceous oxygen demand:

Total Mass Oxygen Required $=$ Mass Oxygen Required for Syntheses

$+$ Mass Oxygen Required for Endogenous Respiration

(6)

Written symbolically, Equation 6 becomes:

$$R_c = a'Q(S_o - S) + b'VX \qquad (7)$$

where

a' $=$ a coefficient dependent on process conditions, expressed here as mass oxygen required per mass total carbonaceous oxygen demand satisfied, M/M

b' $=$ an endogenous oxygen demand coefficient, also dependent on process conditions, expressed here as mass oxygen required/time/mass microbial mass under aeration, M/t/M

X $=$ biomass in aerated zone, M/L^3

It can be shown that Equation 7 is identical to Equation 4 where

$$a' = 1 - \beta Y \text{ and } b' = \beta b \qquad (8)$$

A more sophisticated approach to estimating oxygen demand in aeration systems may be obtained from process models which subdivide the wastewater and mixed liquor into several components. This approach is typified by the comprehensive model developed by the International Association of Water Pollution Research and Control (IAWPRC).[7] The IAWPRC model includes carbon oxidation, nitrification, and denitrification and can predict oxygen utilization for all the processes. This model requires use of a personal computer and dedicated software. Bidstrup and Grady[8] describe such software with capability to model an activated sludge process by up to nine reactors-in-series.

Accurate prediction of oxygen utilization by any approach requires that the values of the model parameters be established for the wastewater under study. Equation 4 involves two model parameter values, Y and b, while the IAWPRC model may involve up to 11 parameters which must be determined.[8] For preliminary studies, literature values of the parameters may suffice; but, for design of large facilities, pilot studies may be warranted.

Nitrogenous Oxygen Demand. Nitrogenous oxygen demand is the result of the oxidation of ammonia nitrogen to nitrate nitrogen. Oxygen demand is typically calculated on a stoichiometric basis, theoretically equal to 4.57 kg oxygen per kg of ammonia converted.[9] Actually, because of the synthesis of some ammonia in the process, the oxygen requirement is closer to 4.2 kg oxygen/kg nitrate-nitrogen produced. However, the stoichiometric figure is often used in engineering calculations for aeration requirements.[9] Nitrogen entering the aeration basin usually is in the form of organic nitrogen, and is often described as total Kjeldahl nitrogen (TKN). Although some of the organic nitrogen is nonbiodegradable, the remainder may be converted to ammonia for synthesis or oxidation to nitrate or nitrite. It is the ammonia nitrogen that is oxidized that is of concern in estimating oxygen demand.

The nitrogen available for nitrification may be calculated in several ways. Accurate estimates can be determined by quantifying the mass of nitrogen in each pathway.[10] Less precise methods can be used where data are not available to effectively elicit each pathway. For example, the mass of waste sludge nitrogen may be subtracted from the mass of influent TKN to estimate nitrogen available for oxidation. A further simplification assumes that the influent ammonia nitrogen represents the nitrogen available for oxidation.

It should be emphasized that process conditions will control biological nitrification. Temperature, pH, aeration basin DO, and θ_c are among the more important variables controlling the process.[9,10] Even when a plant is not engineered to nitrify, process conditions may occur that will result in nitrification. This can lead to oxygen demands that exceed capacity resulting in low–to–zero dissolved oxygen concentrations.

Under appropriate conditions, nitrate-nitrogen may undergo denitrification. The process of denitrification, resulting in the production of gaseous nitrogen end-products, may significantly reduce the oxygen demand of the system. In this case, nitrate-nitrogen serves as an electron acceptor, in much the same way as does oxygen, satisfying a portion of the oxygen demand. A stoichiometric calculation of this pathway indicates that 2.86 kg oxygen demand is satisfied per kg nitrate-nitrogen denitrified.[9,10] This "denitrification credit" may be used in the overall calculation of total oxygen demand. It should be used with caution, however. The process design must ensure that denitrification will occur dependably throughout the design life of the facility. Often times this "credit" is omitted in the calculations to provide a degree of conservatism in the aeration system design.

The net nitrogenous oxygen demand may be estimated by the following equation:

$$R_n = 4.57 \ Q(N_o - N) - 2.86 \ Q(N_o - N - NO_3) \qquad (9)$$

where N_o, N may be expressed as the oxidizable nitrogen in the influent and effluent of the process (M/L^3) and NO_3 is the effluent nitrate nitrogen (M/L^3).

As described earlier, models are available to provide more precise estimates of the kinetics of biological processes including nitrification and denitrification.[7] These models require values for up to 19 kinetic and stoichiometric parameters and are useful in predicting the extent and rate of nitrificiation and denitrification. Software associated with these models[8] is especially useful for estimating spatial and temporal variation in oxygen demands.

Inorganic Chemical Oxygen Demand. Oxygen demand may also occur as the result of oxidation-reduction reactions because of the presence of certain reduced compounds within the wastewater. This oxygen demand is most often estimated based on a stoichiometric calculation for the given reaction. For example, oxidation of hydrogen sulfide by oxygen may be written by the following equation.

$$H_2S + 2O_2 \rightarrow H_2SO_4 \qquad (10)$$

In this example, 1.88 kg of oxygen are required to oxidize each kg of H_2S [$(2 \times 32)/(1 \times 34)$]. Similar calculations can be made for other reduced compounds present in wastewaters in significant quantities.

Another approach to estimating inorganic oxygen demands is through experimental measurement of oxygen uptake over a short interval of time, often referred to as the immediate oxygen demand (IDOD). Although at one time included in the publication "Standard Methods for Examination of Water and Wastewater," this method has been recently dropped. Among reasons for its elimination were the arbitrary use of 15 minutes as a test period, inaccuracy of the measurement, and the possible occurrence of an "iodine demand" rather than an oxygen demand. Caution should be used in the interpretation of laboratory measurements.

Example Calculations. The following two examples demonstrate the calculation of carbonaceous oxygen demand and carbonaceous-plus-nitrogenous oxygen demand.

Example 1

Estimation of Carbonaceous Oxygen Demand

> Given: Average wastewater flow   2.0 mgd
> Average settled BOD_5 200 mg/L
> Ratio ultimate BOD/BOD_5 1.5
> Effluent Requirements: BOD_5 20 mg/L
>   VSS   20 mg/L
> Biodegradable effluent suspended solids = 65%
> θ_c = 5 days
> β = 1.42 mg O_2/mg VSS
> Y = 0.37 mg VSS/mg ultimate BOD*
> b = 0.05 day^{-1}*

where the asterisk (*) indicates these values must be determined from the literature or from laboratory–or full-scale tests.

A. Using Equation (4):
   i. Estimate effluent soluble ultimate BOD:
      Biodegradable portion of effluent VSS = 0.65 × 20 mg/L = 13 mg/L
      Total carbonaceous BOD of VSS = 1.42 × 13 mg/L = 18.5 mg/L
      Ultimate BOD of effluent required = 20 mg/L × 1.5 = 30 mg/L
      Effluent soluble ultimate BOD = 30 − 18.5 = <u>11.5</u> mg/L
   ii. Estimate influent ultimate BOD
      Influent ultimate BOD = 200 mg/L × 1.5 = 300 mg/L
   iii. Calculate R_c from Equation (4):

$$R_c = Q(S_o - S)\left(\frac{1 + b\theta_c - \beta Y}{1 + b\theta_c}\right)$$

$$= 2\,mgd \times 8.34(300 - 11.5)\left(\frac{1 + .05 \times 5 - 1.42 \times .37}{1 + .05 \times 5}\right)$$

$$R_c = \underline{2791}\ lb/day\ (1270\ kg/d)$$

B. Using Equation (6)
   i. Estimate a′ and b′
      $a' = 1 - \beta Y = 1 - 1.42 \times .37$
              $= 0.4746$
      $b' = \beta b = 1.42 \times .05 = 0.071$

   ii. Estimate VX:
      From Reference 4,

$$VX = \frac{Q\theta_c Y(S_o - S)}{1 + b\theta_c}$$

$$VX = \frac{2\ mgd \times 8.34 \times 5 \times .37(300 - 11.5)}{1 + .05 \times 5}$$

$$= \underline{7122}\ lb\ (3237\ kg)\ VSS$$

   iii. Calculate R_c from Equation (6)
      $R_c = a'Q(S_o - S) + b'XV$
        $= .4746 \times 2\ mgd \times 8.34(300 - 11.5) + .071(7122)$
        $= \underline{2791}\ lb/day\ (1270\ kg/d)$

Example 2

Estimation of Carbonaceous plus Nitrogenous Oxygen Demand

Given: Same as Example 1 with following additions/changes:
Change: $\theta_c = 10$ days
Effluent $NH_4 \leq 1.0$ mg/L N
Addition: Raw Wastewater TKN = 41 mg/L-N
Primary Effluent TKN = 36 mg/L-N
Waste Sludge TKN = 225 lb/day (102 kg/d) [83 lb/d in primary sludge and 142 lb/d in waste-activated sludge]
Anoxic Zone Designed to Provide Dentrification
Effluent NO_3-N = 5 mg/L
NH_4-N = 1 mg/L

i. Estimate nitrogen available for nitrification:
Raw wastewater TKN = 41 mg/L $\times$ 2.0 mgd $\times$ 8.34
= 684 lb/day (321 kg/d)
Waste sludge TKN = 225 lb/day (102 kg/d)
Nitrogen available for nitrifica-
tion = 684 − 225 = 459 lb/day (209 kg/d)

ii. Estimate nitrogen nitrified:
Nitrogen nitrified = Available Nitrogen − NH_4–N effluent
= 459 lb/d − 1 $\times$ 2.0 $\times$ 8.34 = 442 lb/day (201 kg/d)

iii. Mass nitrate-in effluent
NO_3–N = 5.0 $\times$ 2 $\times$ 8.34 = 83 lb/day (38 kg/d)

iv. Calculate nitrogenous oxygen demand:
$$R_n = 4.57 \, Q(N_o\text{–}N) - 2.86 \, Q(N_o\text{–}N - NO_3)$$
$$= 4.57 \times 442 - 2.86(442 - 83)$$
$$= 2020 - 1027$$
$$R_n = \underline{993} \text{ lb/day (451 kg/d)}$$

v. Calculate carbonaceous plus nitrogenous demand:

$$R_T = R_c + R_n$$

$$= Q(S_o - S) \left(\frac{1 + b\theta_c - \beta Y}{1 + b\theta_c} \right) + R_n$$

$$= 2 \times 8.34(300 - 11.5) \left(\frac{1 + .05 \times 10 - 1.42 \times .37}{1 + .05 \times 10} \right) + 933$$

$$= 3126 + 933$$

$$R_T = \underline{4059} \text{ lb/day (1845 kg/d)}$$

vi. Alternative calculation for R_n
Assume influent TKN to aeration tank is all available for nitrification. (A more conservative estimate.)
$$R_n = 4.57 \, Q(N_o - N) - 2.86(N_o - N - NO_3)$$
$$= 4.57 \times 2 \times 8.34(36 - 1) - 2.86(36 - 1 - 5) \times 2 \times 8.34$$
$$= 2668 - 1431$$
$$= \underline{1237} \text{ lb/day (vs. 993 lb/day) (562 vs 451 kg/d)}$$

Variations in Process Oxygen Requirements. The estimates of process oxygen requirements described in Examples 1 and 2 predict a spatial and temporal average, pseudo steady-state oxygen demand in the aeration tank. However, the design engineer must also have information on the magnitude of variation of this demand with time and within the reaction vessel so that sufficient aeration capacity and flexibility can be designed and installed. Sufficient capacity must be installed in each portion of the system to insure that adequate oxygen is present under peak demands. However, sufficient turndown capability must also be provided to allow efficient operation at reduced demands (Figure 2-1).[11]

Temporal variations will occur because of variation in pollutional load to the aeration tank. These variations may be estimated from statistical analysis of data collected on wastewater loading (carbonaceous and nitrogenous) to the reactor.[3] This information may be used in Equations 4, 7, and 9 to provide estimates of peak, minimum, and average oxygen demands at a given point within the system. In the absence of actual wastewater load variations, rules of thumb based on experience with similar wastewaters may be used to estimate peak and minimum demands.

Spatial variations in oxygen demand are dependent on the kinetic relationships between growth rates of the biomass and substrate and dissolved oxygen concentrations. They are also dependent on the hydraulic regime of the process (e.g., plug flow, tanks-in-series, sludge, and recycle flows). In plug flow tanks, the demand for oxygen will be greater at the inlet end where the availability of substrate does not limit the rate of oxidation compared to that at the outlet end where the rate of oxidation is typically limited because of substrate depletion. Good estimates of this variation can be obtained by use of process models such as the IAWPRC model adapted to flexible software packages[7,8] described earlier. While more precision is possible with this approach, these models can be used only if they have been calibrated to the specific wastewater and system being designed.

There is a more empirical approach to estimate spatial variations in oxygen demand. Figure 2-2 illustrates one example of spatial variation in oxygen demand along a plug flow tank.[12] Data in the U.K. have indicated that the rate of oxygen demand for nitrification is constant from inlet to outlet of a plug flow system provided that dissolved oxygen concentrations are above 0.5 mg/L.[13,14] For carbonaceous demand only, spatial distribution in oxygen demand for long, narrow tanks (L/W > 20) typically designed in the U.K. are given in Table 2-1. Assuming uniform nitrification along the entire basin, the spatial distri-

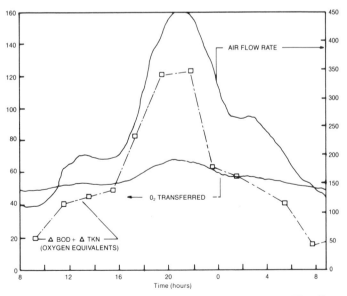

FIGURE 2.1—Relationship Between Oxygen Transfer, Air Flow Rates, and Carbonaceous and Nitrogenous Oxygen Demands Under Diurnal Loading Conditions.

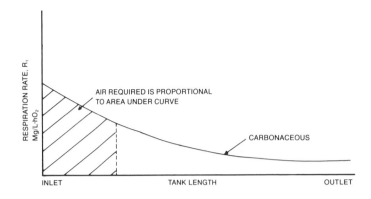

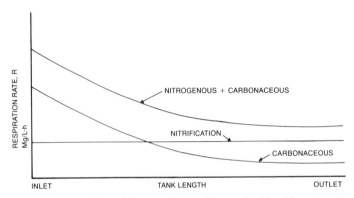

FIGURE 2.2—Typical Oxygen Uptake Curves for Plug Flow Aeration Tanks.

bution attenuates that demand as shown in Table 2-1. Typical spatial distributions used in U.S. practice may differ from Table 2-1 values depending on tank geometry and recirculation ratios.

Table 2-1   Variation in Proportion of Oxygen Demand Along the Length of a Plug Flow Aeration Tank (L/W > 20) (12)

Proportion of Aeration Tank Volume (%)	CARBONACEOUS DEMAND		CARBONACEOUS + NITROGENOUS DEMAND	
	Proportion of Demand (%)	Diurnal Range (%)	Proportion of Demand (%)	Diurnal Range (%)
20	60	40–85	46	33–62
20	15	5–20	17	10–20
20	10	5–15	14	10–17
20	10	5–15	13	10–16
20	5	<1–10	10	7–13

*M*ixing

A secondary function of aeration devices in aerobic biological systems has been to furnish sufficient energy for mixing. Ideally, mixing energy should be large enough to insure that dissolved substrate and oxygen are well dispersed throughout a given section of the volume and that solids remain suspended within the reactor. It does not necessarily mean that both soluble and suspended material are uniform throughout the aerated volume, however (e.g., plug flow aeration tank, looped tubular reactor with a point source of oxygen).

It should be noted that the power required for oxygen demand depends on both substrate concentration and volume whereas power for mixing depends only on reactor volume. There is a point, depending on substrate concentration, where power requirements for mixing exceed those for oxygen transfer. In high biomass systems, it is typical that oxygen demand will control power requirements whereas aerated lagoon power requirements are often dictated by mixing energy.

Detailed discussions of mixing energy by both diffused air and mechanical air systems are found in Chapters 4 and 5.

R*eferences*

1. *Recommended Standards for Sewage Works*, Great Lakes—Upper Mississippi River Board of State Sanitary Engineers (1978 Edition).
2. Fair, G.M. and Geyer, J.C., *Water Supply and Wastewater Disposal*, Wiley and Sons, New York (1954).
3. *Wastewater Treatment Plant Design*, Water Pollution Control Federation Manual of Practice 8, WPCF, Washington, D.C. (1977).
4. Lawrence, A.W. and McCarty, P.L., "Unified Basis for Biological Treatment Design and Operations," Jour. San. Engr. Div., ASCE, 96, 757 (1970).
5. Eckenfelder, W.W. Jr., *Principles of Water Quality Management*, CBI Publishing Company, Inc., Boston (1980).
6. Metcalf and Eddy, Inc., *Wastewater Engineering: Treatment, Disposal, Reuse*, 2nd Edition, McGraw-Hill Book Co., New York (1979).
7. Grady, C.P.L. Jr., Guyer, W., Henz, M., Marias, G.V.R., and Matsuo, T., "A Model for Single Sludge Wastewater Treatment Systems," *Water Science and Technology*, 18, 47 (1986).
8. Bidstrup, S.M. and Grady, C.P.L. Jr., "SSSP-Simulation of Single Sludge Processes," *J. WPCF*, 60, 351 (1988).
9. U.S. EPA, *Process Design Manual for Nitrogen Removal* (October 1975).
10. Water Research Commission, *Theory, Design, and Operation of Nutrient Removal Activated Sludge Processes*, Pretoria, South Africa (1984).
11. Stephenson, J.P., "Automatic Dissolved Oxygen Control in the Activated Sludge Process," In *Proceedings: Seminar Workshop on Aeration System Design, Testing, Operation, and Control*, Ed. W.C. Boyle, EPA 600/9-85-005, 201 (January 1985).
12. Boon, A.G. and Chambers, B., "Design Protocol for Aeration Systems—U.K. Perspective," In *Proceedings: Seminar Workshop on Aeration System Design, Testing, Operation, and Control*, Ed. W.C. Boyle, EPA 600/9-85-005, 99 (January 1985).
13. Downing, A.L., Jones, K., and Hopwood, A.P., "Some Factors of Importance in the Design of Activated Sludge Plants," In *Joint Symposium on New Chemical Engineering Problems in the Utilization of Water*, Amer. Inst. Chem. Engr. and Institution of Chemical Engineers (1965).
14. Downing, A.L., Boon, A.G., and Bayley, R.W., "Aeration and Biological Oxidation in the Activated Sludge Process," *J. Proc. Inst. Sew. Purif.*, 66 (1962).

Chapter 3
Oxygen Transfer Modeling

THEORETICAL DEVELOPMENT

The basic model for oxygen transfer in a dispersed gas-liquid system is given.[1]

$$\text{Rate of Mass Transfer per Unit Volume of Liquid} = \text{Volumetric Mass Transfer Coefficient} \times \text{Driving Force} \quad (1)$$

Expressed in terms of the overall liquid-phase, volumetric mass transfer coefficient, this relationship becomes

$$\hat{W} = \hat{K}_L a^* (\hat{C}^* - \hat{C}) \qquad (2)$$

where:

the superscript ^ is employed to indicate the parameters at a particular location within the volume of liquid being aerated, and

W = mass transfer rate per unit volume of liquid, $ML^{-3}t^{-1}$

$K_L a^*$ = overall volumetric mass transfer coefficient in clean water based on the liquid film resistance, t^{-1}

C^* = dissolved oxygen saturation concentration, ML^{-3}

C = dissolved oxygen concentration, ML^{-3}

the symbols M, L, and t denote dimensions of mass, length, and time, respectively.

Equation 2 applies only to a particular point in an oxygen transfer system. In general, the total rate of oxygen transfer is of interest and could be evaluated by integrating the local transfer rate over the liquid volume. For real aeration systems, however, spatial variations in $K_L a$, C^* and C are not well-defined and this integration is not practical. Instead, the total transfer rate is conventionally expressed in terms of an apparent average volumetric mass transfer coefficient and oxygen saturation concentration as,

$$OTR = K_L a \, V \, (C^*_\infty - C) \qquad (3)$$

where:

OTR = oxygen transfer rate occurring in the liquid volume, V, Mt^{-1}

$K_L a$ = apparent spatial average volumetric mass transfer coefficient in clean water, t^{-1}

V   = liquid volume of the aeration system, L^3
C^{*}_{∞} = spatial average dissolved oxygen saturation concentration
     approached at infinite aeration time, ML^{-3}
C   = spatial average dissolved oxygen concentration, ML^{-3}

The value of the saturation concentration, C^{*}_{∞}, to be applied in
Equation 3 is the spatial average of the values determined from a clean
water performance test[2] and is not corrected for gas-side oxygen depletion. Consequently, the $K_L a$ value used here is an apparent value
because it is determined based on the uncorrected value of the saturation concentration. An alternate approach, based on a saturation concentration corrected for gas-side oxygen depletion and true volumetric
mass transfer coefficient, can also be applied to this situation. Brown
and Baillod[3] have shown that identical results are obtained using either
approach. The approach based on the uncorrected value denoted by
C^{*}_{∞} is advantageous because:

C^{*}_{∞} does not depend on the transfer rate, and
C^{*}_{∞} can be accurately and precisely measured for a given aeration
    system using clean water tests.

Alternatively, C^{*}_{∞} can be estimated for design purposes from the surface saturation concentration and effective saturation depth d_e by

$$C^{*}_{\infty} = C^{*}_{ST} \left(\frac{p_b - p_{vt} + \gamma_w d_e}{p_s - p_{vt}} \right) \qquad (4)$$

where:
C^{*}_s = tabular value of DO surface saturation concentration
    (ML^{-3}), at temperature (T), standard total pressure
    $(p_s = 1.00$ atm.$)$, and 100% relative humidity.
C^{*}_{∞} = spatial average dissolved oxygen saturation concentration
    approached at infinite aeration time at temperature (T),
    ML^{-3}
d_e = effective saturation depth, L
p_s = standard total pressure (1.00 atmosphere of air at 100% rel-
    ative humidity), $ML^{-1}t^{-2}$
p_b = atmospheric pressure, $ML^{-1}t^{-2}$
p_{vt} = saturated vapor pressure of water at temperature (T),
    $ML^{-1}t^{-2}$
γ_{wT} = mass density of water at the temperature (T), $ML^{-2}t^{-2}$

The effective saturation depth, d_e, represents the depth of water
under which the total pressure (hydrostatic plus atmospheric) would
produce a saturation concentration equal to C^{*}_{∞} for water in contact
with air at 100% relative humidity. This can be calculated using Equation 4 based on a spatially averaged value of $C^{*}_{\infty T}$ measured by a clean
water test. For design purposes, d_e can be estimated from clean water
test results on similar systems; it ranges from 5 to 50% of tank liquid
depth. Table 3.1 summarizes effective saturation depths reported for a
study of a wide variety of aeration systems.[4]
    The model given by Equation 3 accurately and precisely describes the
rate of oxygen transfer in a wide variety of aeration systems.[4] This
model can be applied directly based on parameters determined in the
clean water test (see Chapter 8).

Table 3.1. Summary of effective saturation depth for various aeration systems.

Aeration System	Effective Saturation Depth (d_e), as a Percentage of Tank Liquid Depth
Coarse bubble diffused air	26 to 34% of depth
Fine bubble diffused air	21 to 44% of depth
Low speed surface aeration	5 to 7% of depth

*A*PPLICATION OF CLEAN WATER OXYGEN TRANSFER COEFFICIENTS

Aeration system performance information is generally available for clean water conditions and is reported for standard temperature and barometric pressure. The design process generally requires that the oxygen transfer rate be determined from process and load information as described in Chapter 2. Consequently, the actual oxygen transfer rate delivered by an aeration system under process conditions must be determined by correcting the standard clean water performance information.

CORRECTION FACTORS FOR PROCESS WATER CHARACTERISTICS—TEMPERATURE AND PRESSURE

Using standard clean water test parameters in the design or evaluation of process systems requires correcting for the influence of process water characteristics, temperature, and pressure on the oxygen transfer coefficient and oxygen saturation value. This correction is accomplished by using the factors, as summarized in Table 3.2: alpha, beta, theta, tau, and omega. Each of these factors is generally expressed as a ratio of the corrected parameter at process conditions to the same parameter at standard clean water conditions.

Table 3.2. Summary of correction factors for oxygen transfer coefficient and oxygen saturation value.

Correction Factor	Accounts for Effect Of	On
alpha (α)	Process Water Characteristics	Transfer Coefficient ($K_L a$)
beta (β)	Process Water Characteristics	Saturation Concentration (C^*_∞)
theta (θ)	Temperature	Transfer Coefficient ($K_L a$)
tau (τ)	Temperature	Saturation Concentration (C^*_∞)
omega (Ω)	Pressure	Saturation Concentration (C^*_∞)

EFFECT OF WATER CHARACTERISTICS—ALPHA AND BETA FACTORS The effect of process water characteristics is commonly accounted for by the alpha and beta factors, defined by

$$\alpha = \frac{\text{process water } K_L a}{\text{clean water } K_L a} = \frac{K_L a_f}{K_L a} \tag{5}$$

$$\beta = \frac{\text{process water } C^*_\infty}{\text{clean water } C^*_\infty} = \frac{C^*_{\infty f}}{C^*_\infty} \tag{6}$$

in which the subscript f denotes the process water condition. It is reasonable to expect that the influence of process water characteristics on the various saturation values will be similar, so that β can be expressed in terms of tabulated surface saturation values as:

$$\beta = \frac{C^*_{\infty f}}{C^*_\infty} = \frac{C^*_{sf}}{C^*_s} \tag{7}$$

in which the subscript s denotes surface saturation values.

The alpha factor ranges from approximately 0.2 to greater than 1.0.[5,6] It is influenced by many process conditions including surfactants, turbulence, power input per unit of volume, tank geometry, geometric scales between aeration tank and aeration device, bubble size, degree of treatment, and other wastewater characteristics. Ideally, the alpha factor would be measured by conducting full-scale oxygen transfer tests with clean water and process water, but normally this is impractical.

Several studies have described small-scale (less than 200 liter) oxygen transfer tests for measurement of the alpha factor.[5] Alpha factor measurement, however, in small scale vessels is only an educated guess at best. In selecting an alpha factor, one should remember that, for a given wastewater stream, the alpha factor is normally not constant; therefore, a possible range of alpha values should be considered in estimating the transfer rate under process conditions.[6,7]

The beta factor can vary from approximately 0.8 to 1.0 and is generally close to 1.0 for municipal wastewaters. Because it cannot be measured by a membrane probe and because many wastewaters contain substances that interfere with the Winkler method[8] for DO determination, beta is difficult to measure accurately. The value of beta, therefore, should be based on the dissolved solids content of the process water and should be calculated as the ratio of the DO surface saturation concentration in water, with dissolved solids equal to that of the process water, to the DO surface saturation concentration in clean water. The corresponding surface saturation concentrations can be determined from published tabular values of these concentrations as a function of dissolved solids concentration or chlorinity.[6]

EFFECT OF TEMPERATURE—THETA AND TAU FACTORS

The influence of temperature on the oxygen transfer coefficient and oxygen saturation value can be expressed in terms of the factors theta and tau, defined by

$$\theta^{(T-20)} = \frac{K_L a_T}{K_L a_{20}} \tag{8}$$

$$\tau = \frac{C^*_{\infty T}}{C^*_{\infty 20}} \tag{9}$$

in which the subscript T denotes process temperature and 20 denotes the standard temperature of 20 degrees Celsius. The influence of temperature on the various oxygen saturation concentrations will be similar. Therefore, τ can be conveniently calculated based on published DO surface saturation values:

$$\tau = \frac{C^*_{\infty T}}{C^*_{\infty 20}} = \frac{C^*_{sT}}{C^*_{s\,20}} \tag{10}$$

Values of theta reported in the literature have ranged from 1.008 to 1.047 and are influenced by geometry, turbulence level, and type of aeration device.[5] Because there is little consensus regarding the accurate prediction of theta values, clean water testing for the determination of OTR values should be at temperatures close to the design application temperature. The clean water test standard[2] recommends that the value of theta be taken equal to 1.024 unless experimental data for the particular aeration system indicate conclusively that the value is significantly different from 1.024.

EFFECT OF PRESSURE—OMEGA FACTOR

It is necessary to correct C^*_∞ for differences in atmospheric pressure between the test, standard, and process conditions. The value of C^*_∞ is not a linear function of total atmospheric pressure, but is a linear function of the partial pressure of dry air at the saturation depth. Therefore, the pressure correction factor, Ω, is defined as

$$\Omega = \frac{C^*_\infty \text{ at } p_b}{C^*_\infty \text{ at } p_s} = \frac{P_b + \gamma_{wT} d_e - p_{vT}}{p_s + \gamma_{wT} d_e - p_{vT}} \tag{11}$$

where the symbols are as defined earlier. When tank depths are less than 6 meters, Ω may be approximated by p_b/p_s.

APPLICATION TO PROCESS CONDITIONS

The details of the method employed to apply clean water test results to process conditions depend on the model used to analyze the clean water data. The procedure given here presumes that the Clean Water Test Standard[2] has been applied to give representative average values of K_La, c^*_∞, and d_e for clean water.

The transfer rate at process conditions, OTR_f, can be expressed in terms of the standardized K_La_{20} and $C^*_{\infty 20}$ using the correction factors, α, β, θ, τ, and Ω, as,

$$OTR_f = K_La_{20} \, \alpha \, \theta^{(T-20)} \, (\Omega \tau \beta C^*_{\infty 20} - C) \, V \qquad (12)$$

where:

T = temperature of the process water
C = spatial average DO concentration over the process water volume, and

the other terms are as defined previously.

Aeration design equations are frequently expressed in terms of the Standardized Oxygen Transfer Rate (SOTR), defined as the rate of oxygen transfer in clean water at zero DO and a specified temperature, usually 20° C.

$$SOTR = K_La_{20} \, C^*_{\infty 20} V \qquad (13)$$

The SOTR is a hypothetical value based on zero DO in the aeration zone; this condition is not usually attainable in real aeration systems operating in process water.

Equations 12 and 13 can be combined to give the Process Oxygen Transfer Rate (OTR_f) in terms of the SOTR and the correction factors as

$$OTR_f = \frac{\alpha(SOTR) \, \theta^{(T-20)}}{C^*_{\infty 20}} \, (\tau \, \beta \, \Omega \, C^*_{\infty 20} - C) \qquad (14)$$

A guide to the application of Equation 14 is found in Table 3.3, which indicates the source of information for the parameters needed to estimate the OTR_f. The value of $C^*_{\infty 20}$ must be either known from clean water measurements or estimated based on a reasonable value of d_e, according to Equation 4. The average DO value (C) must be determined from the process water conditions and should be evaluated as the process level DO concentration averaged over the entire aeration volume. It should not be taken as the DO concentration in the influent to a point source aerator. Rather, it should represent the average DO over the whole aeration volume.[3]

The appropriate average process DO concentration for design depends on process characteristics. A high DO value will require an aeration system having a large SOTR and high-power requirements, whereas a low DO value may yield poor performance. Available evidence indicates that at least 1 mg/L of DO is required for satisfactory functioning of aerobic metabolism in activated sludge. There are indications, however, that significantly greater values may help to control filamentous organisms responsible for bulking sludge. Moreover, certain process modifications for denitrification may require lower DO values.

Although the application of clean water SOTR values to estimate transfer rates in process water is conceptually straightforward, the estimate of OTR_f is subject to considerable uncertainty because the correction factors are not exact, particularly in the α value. This uncertainty is magnified when the process water application is based on tank geometry and temperature that differ from those of the clean water test.

Table 3.3. Guide to the application of Equation 14.

Parameter	Source of Information
$C^*_{\infty\,20}$	Clean water test results, or estimated from Equation 4
d_e	Clean water test results, or estimated
SOTR	Clean water test results
C	Given by the process conditions
T	Given by the process conditions
τ	Calculated based on tabulated DO surface saturation values (Equation 10)
Ω	Calculated based on d_e and barometric pressure (Equation 11)
α	Estimated based on experience and on measured $K_L a$ values
β	Calculated based on total dissolved solids measurements
θ	Taken as 1.024 unless experimentally proven to be different

REFERENCES

1. Treybal, R.E., "Mass Transfer Operations" (2nd Ed.). McGraw Hill, New York, p. 95 (1968).
2. "A Standard for the Measurement of Oxygen Transfer in Clean Water." Amer. Soc. of Civ. Eng., New York, N.Y. (1984).
3. Brown, L.C. and Baillod, C.R., "Modeling and Interpreting Oxygen Transfer Data." *J. Env. Eng. Div.*, ASCE, **108**, EE4, 607 (1982).
4. Baillod, C.R., Paulson, W.L., McKeown, J.J., and Campbell, H.J., Jr. "Accuracy and Precision of Plant Scale and Shop Clean Water Oxygen Transfer Tests." *J. Water Pollut. Control Fed.*, **58**, 290 (1986).
5. Stenstrom, M.K. and Gilbert, R.G., "Effects of Alpha, Beta and Theta Factors and Surfactants on Specification Design and Operation of Aeration Systems." *Water Research*, **15**, 643 (1981).
6. Doyle, M.L. and Boyle, W.C., "Translation of Clean to Dirty Water Oxygen Transfer Rates." "Proc. Seminar Workshop on Aeration System Design, Testing, Operation and Control." EPA—600 19—85—45 (Jan. 1985).
7. Huang, H.J. and Stenstrom, M.K., Evaluation of Fine Bubble Alpha Factors in Near Full-Scale Equipment." *J. Water Pollut. Control Fed.*, **57**, 1143 (1985).
8. "Standard Methods for the Examination of Water and Wastewater" (16th Ed.). Water Pollut. Control Fed., Alexandria, Va. (1985).

Chapter 4
Diffused Air Systems

INTRODUCTION

Diffused aeration has been employed in wastewater treatment since the turn of the century.[1] In the early applications, air was introduced through open tubes or perforated pipes located at the bottom of the aeration tank. The desire for greater efficiency led to the development of porous plate diffusers that produce small bubbles and result in higher transfer efficiencies.[2] Porous diffuser plates were used in the activated sludge process as early as 1916 and they became the most popular method of aeration by the 1930s.[3,4,5] Unfortunately, these diffusers experienced serious clogging problems at several installations. Gradually, the use of porous diffusers fell into disfavor, and systems requiring lower maintenance became predominant during the period of relatively cheap energy before 1972.[6] Typically, these devices used fixed orifices (i.e., 0.6 cm or more in diameter) to produce a relatively large bubble. Rapid escalation in power costs since the "energy crisis" in the early 1970s rekindled interest in porous media devices and has resulted in considerable effort to optimize the performance of all types of aeration systems.

Diffused aeration is defined as the injection of gas (air or oxygen) under pressure below the liquid surface. All of the equipment described in this chapter meets this definition. An arbitrary classification, however, is made for hybrid equipment that combines gas injection with mechanical pumping or mixing equipment. Jet aerators, aspirating propeller pumps, and U-tube aerators are classified as diffused aeration equipment, while combination turbine-sparge aerators are classified as mechanical aeration equipment; the latter are discussed in Chapter 5. Equipment whose primary purpose is not the transfer of oxygen (e.g., air lift pumps) are not addressed in this manual.

This chapter presents information on the various types of diffused aeration equipment currently available, including discussion of reported performance characteristics and factors affecting performance. General design considerations associated with diffused aeration systems are also discussed. The latter part of the chapter presents information on blower systems.

Description of Diffused Air Systems

The wastewater treatment industry has witnessed the introduction of a wide variety of air diffusion equipment. In the past, the various devices commonly have been classified as either "fine bubble" or "coarse bubble," a designation that supposedly reflected the device's efficiency in transferring oxygen. Unfortunately, fine bubble diffused aeration is difficult to define and the demarcation between fine and coarse bubbles is not well-differentiated.[7] Also, there has been considerable debate and confusion over which classification applied to certain equipment. For these reasons, the current industry preference is to categorize diffused aeration systems by the physical characteristics of the equipment.

In the discussion of diffused aeration systems that follows, the various equipment has been divided into three categories: porous diffusers; non porous diffusers; and other aeration devices such as jet aerators, aspirating propeller pumps and U-tube aeration. The reader is cautioned not to draw generalities about equipment performance based solely on these labels. These classifications are intended more as a guide for organization than as a categorical statement of performance.

POROUS DIFFUSER SYSTEMS. Use of porous diffusers has gained renewed popularity because of the relatively high oxygen transfer efficiency (OTE) exhibited by most of these systems. An excellent reference on this subject was published by the Environmental Protection Agency in cooperation with the ASCE Committee on Oxygen Transfer; the report is titled *Summary Report: Fine Pore (Fine Bubble) Aeration Systems.*[7] Much of the information presented in this chapter concerning the characteristics and performance of porous media systems was derived from this source, and the reader is encouraged to review the report for further discussion of specific topics.

TYPES OF MEDIA USED. Numerous materials have been used in the manufacture of porous diffusers. These generally fall into the categories of either rigid ceramic and plastic materials or flexible plastic or cloth sheaths.

Ceramic Materials. The oldest and most common type of porous material on the market is the ceramic type. It consists of rounded or irregular-shaped mineral particles bonded together to produce a network of interconnecting passageways through which compressed air flows. As the air emerges from the surface pores, pore size, surface tension, and air-flow rate interact to produce a characteristic bubble size.[8]

Early designs used porous silica plates cast into concrete plenums in the tank floor.[9] Some of these early designs, such as those in Milwaukee, Chicago, Cleveland, and Toronto, are still in operation. Today, the majority of ceramic diffusers being marketed are manufactured from aluminum oxide. Other types of ceramic media composition include vitreous-silicate–bonded grains of pure silica and resin-bonded grains of pure silica.[8]

Plastic Materials. A more recent development is the use of porous plastic materials. As with the ceramics, a material is created that consists of numerous interconnecting channels or pores through which compressed air can pass. Claimed advantages of the plastic material over aluminum oxide are its lighter weight (which makes it especially well suited to lift-out applications), lower cost, better durability, and, depending on the actual material, greater resistance to breakage. Disadvantages include its reduced strength and susceptibility to creep. Porous plastics are made from numerous thermoplastic polymers. The two most common types of plastic materials in use are high-density polyethylene (PE) and styrene-acrylonitrile (SAN).

Flexible Sheaths. Flexible diffusers have been in use for approximately 40 years. Originally known as "sock" diffusers, they were made from materials such as plastic, synthetic fabric cord, or woven cloth. Because of significant fouling problems, there is essentially no current market for the early sock design.

Within the last several years, a new type of flexible diffuser has been introduced. It consists of a thin flexible sheath made from soft plastic or synthetic. Air passages are created by punching minute slots or holes in the sheath material. When the air is turned on, the sheath expands and each slot acts as a variable aperture; the higher the air-flow rate, the greater the opening. The sheath material is supported by a tubular or disc frame. This new generation of flexible diffuser has been in operation at numerous facilities for several years. The new sheath material has reduced the severe fouling problems associated with the earlier woven fabric design, although it may be subject to change over time. There have been cases in which the slits have closed off because of plugging, which resulted in increased pressure loss.

Like porous plastic media, flexible sheaths offer the advantage of light weight. A disadvantage of flexible sheaths is the potential for creep. This phenomenon may enlarge the air slits over time, which produces larger bubbles and reduces oxygen transfer efficiency. The amount of creep is affected by the character of the membrane, the method of making the openings, the geometry of the diffuser, the operating back pressure, and the environmental conditions.

Both disk and tube sheaths vary in thickness, flexibility, and the degree that their performance characteristics change with use. The sheaths must be replaced periodically. Frequencies for replacement have ranged from 2 years to more than 6 years.

TYPES OF POROUS DIFFUSERS. Regarding shape, porous diffusers are available in plates, domes, discs, and tubes (Figures 4.1, 4.2 and 4.3). A brief description of available equipment in each of these shapes is presented in the sections to follow.

Plate Diffusers. Up until the late 1970s, ceramic plates were rivaled in popularity only by tube systems as the most prevalent form of porous media diffuser. Plates are typically 30 cm (12-in.) square and 2.5 to 3.8 cm (1 to 1.5 in.) thick. The plates are installed in the tank by grouting them into recesses in the floor, cementing them into prefabricated holders, or clamping them into metal holders. The metal holders have the potential of fouling associated with corrosion. A chamber underneath the plates acts as an air plenum. The number of plates fixed over a common plenum is not standard and can vary from only a few to more than 500. In current U.S. designs, individual control orifices are not provided on each plate.

Although still available, plate diffusers have declined in popularity since the advent of the dome and disc diffusers. Some possible explanations include (a) problems obtaining uniform air distribution with numerous plates attached to the same plenum, (b) the inconvenience of removing plates when they are grouted in place, and (c) the difficulty in adding diffusers to meet future increases in plant loading.

Plate diffusers can be installed in either a total floor coverage or a spiral roll pattern. Total floor arrangements may include closely spaced rows running either the width (transverse) or length (longitudinal) of the basin or incorporated into a ridge-and-furrow design. Spiral roll arrangements include rows of plates typically located along one or both walls of long narrow tanks.

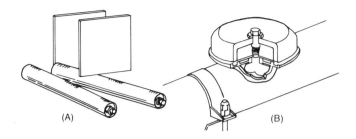

FIGURE 4.1—Typical diffused-air aeration devices. (a) Plate and tube
diffusers. (b) Dome diffuser.

Dome Diffusers. The ceramic dome diffuser was first developed in 1954
and was refined to its present form by 1961. It has become an accepted
standard in England and is enjoying increased popularity in the U.S.
The dome diffuser is essentially a circular disc with a downward-turned
edge. Currently, these diffusers are 18 cm (7 in.) in diameter and
3.8 cm (1.5 in.) high. The media is approximately 15 mm (0.6 in.)
thick on the edges and 19 mm (0.75 in.) on the top or flat surface.
Domes presently are being made only of aluminum-bearing compounds.

The dome diffuser is typically mounted on PVC saddle-type base
plates that are solvent welded to the air distribution piping at the factory. The diffuser is attached to the base plate by a bolt through the
center of the dome. This bolt can be made from numerous materials
including brass, plastic, monel, or stainless steel. Care must be taken
when installing the dome to prevent excessive tightening of the center
bolt, which can cause damage to the diffuser or subsequent air leakage.
A soft gasket of appropriate material is placed between the diffuser and
the base plate. A washer and gasket are also used between the bolt
head and the top of the diffuser. For the air to be better distributed
throughout the system, control orifices are placed in each diffuser
assembly to create additional headloss and to balance the air-flow. In
one design, the fastening bolt is hollowed out and a small hole is drilled
in the side. In another design, the orifice is drilled in the base of the
saddle. Dome diffusers are usually designed to operate at 0.5 L/s
(1 scfm) with a range of 0.25 to 1.0 L/s (0.5–2 scfm). Greater unit air-
flow rates are attainable, but with both a resultant loss in transfer efficiency and an increase in pressure loss.

Domes are generallly installed in a total floor coverage or grid pattern. In some cases in which oxygen demand is low and mixing may
control the design (i.e., near the end of long narrow tanks), the diffusers can be placed in tightly spaced rows along the side or middle of the
basin to create a single-spiral roll or center roll-mixing pattern, respectively. The diffusers are usually mounted as close to the tank floor as
possible, typically within 23 cm (9 in.) of the floor's highest point. The
submergence in some retrofit applications, however, may be controlled
by the available blower discharge pressure.

The air distribution network for dome diffusers is generally constructed from PVC pipe that is supported from the floor using pipe
supports. These supports, typically constructed from stainless steel or
PVC, are adjustable to compensate for the variations in the tank from
elevation. Care must be taken in the design and installation of air
distribution piping to provide for expansion and contraction. With a
dome system or any other system, selection of the proper materials is
very important. If gas cleaning is employed, materials should be compatible with the type of gas used.

Disc Diffusers. Disc diffusers are a relatively recent development. Discs are relatively flat and they are differentiated from the dome diffuser because they do not include a downward-turned, peripheral edge. Although the dome design is relatively standard, current disc diffusers differ in size, shape, method of attachment, and type of material. Most of the discussion that follows pertains to disc diffusers that are constructed using rigid porous media; however, the last paragraph describes a newer type of disc diffuser that employs a flexible membrane.

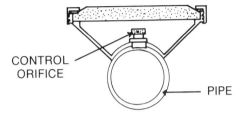

FIGURE 4.2—Typical disc diffuser cross-section.

Disc diffusers are available in diameters that range from approximately 18 to 24 cm (7 to 9.5 in.) and thicknesses of 13 to 19 mm (0.5 to 0.75 in.). With the exception of one design, all discs consist of two flat parallel surfaces. The exception has a raised ring sloping slightly downward toward both the outer edge and the center of the disc. Theoretically, the nonuniform profile aids in producing uniform air flow across the entire disc surface.[10] Although the majority of disc diffusers are made from aluminum oxide, a porous PE disc is also available. One manufacturer offers an interchangable design that allows the ceramic disc to be replaced with a membrane disc and vice versa. Like the dome diffusers, the disc is mounted on a plastic or stainless steel saddle-type base plate. Two basic methods are used to secure the media to the holder: a center bolt or a peripheral-clamping ring. The center bolt method is similar to that used with the domes. The more common method of attaching the disc to the holder is to use a screw-on retainer ring. With this approach, numerous other gasket arrangements are used.

There are two methods of attaching disc diffusers to the air piping. The first is to solvent weld the base plate to the PVC header before shipment to the job site. The second disc diffuser attachment method uses mechanical means of attachment; either a bayonet-type holder is forced into a saddle on the pipe or a wedge section is placed around the pipe and clamps the holder to the pipe. The second approach makes shipping somewhat easier and less bulky making installation of additional future diffusers easier. Unlike the factory mounted solvent welded holders, however, there is an increased possibility of installed holders not being within a common horizontal plane, resulting in poor air distribution.

Disc diffuser assemblies also include individual control orifices in each assembly and have a design air-flow range of 0.25 to 1.5 L/s (0.5 to 3 scfm). The most economical operating range is somewhat dependent on diffuser size. The 18 cm (7 in.) diameter discs are usually operated in the range of 0.25 to 1 L/s (0.5 to 2 scfm), which is similar to the dome diffusers. For the larger discs, with diameters of 22 to 24 cm (8.5 to 9.5 in.), typical lower and upper limits are 0.3 to 0.45 L/s (0.6 to 0.9 scfm) and 1.25 to 1.5 L/s (2.5 to 3 scfm), respectively. In those applications where operation above 1 L/s (2 scfm) is desirable, the control orifice should be sized accordingly so that the headloss produced does not adversely affect the economics of the system. The diffuser layouts for disc diffusers are essentially identical to those used for dome systems. The number of diffusers required, however, may vary depending on the size of the unit and the diffuser media used.

Recently, several types of flexible membrane diffusers have appeared in the market place. These units have diameters of up to 520 mm (20.5 in.) and operate at air-flow rates of 1.5 to 10 L/s (3 to 20 scfm.). The diffuser membrane is typically attached to a plastic base using both a center bolt and a clamp around the periphery. One design uses a wire clamp while another uses a stainless steel band. A third design has the flexible membrane vulcanized to a ring that is attached to the base with stainless screws. The diffusers are typically mounted in full-floor coverage, although other layouts may also be used. Several of these devices are designed in a manner such that, on loss of air pressure, the membrane sits on and seals the orifice to prevent liquid backflow into the distribution piping.

Tube Diffusers. Most of the tube diffusers on the market are of the same general shape. Typically, the media portion is 50 to 60 cm (20 to 24 in.) long and has an outside diameter of 6.4 to 7.6 cm (2.5 to 3.0 in.). The thickness of the media varies. Flexible sheaths of the softened plastic design are very thin, commonly in the range of 0.5 to 3.8 mm (0.02 to 0.15 in.). Synthetic rubber sheaths are substanially thicker. The PE media is usually supplied with a thickness of 6.4 mm (0.25 in.); the SAN media is approximately 15.2 mm (0.6 in.); and the fused ceramic material is in the range of 9.5 to 12.7 mm (0.38 to 0.5 in.).

The holder designs for the ceramic and porous plastic media are very similar. Most consist of two end caps held together by a connecting rod through the center. Gaskets are placed between the media and the end caps to provide an airtight seal. In some cases, a gasket or O-ring is also used in conjunction with the retaining bolt. For the flexible sheath diffusers, the end caps and support frame are one piece. The sheath is installed over the support frame and clamped on both ends. In this design, no gaskets are required. All components of the various tube assemblies are either stainless steel or durable plastic to prevent corrosion. The gaskets are usually of a soft rubber material. Many of these devices incorporate a flapper or other means of sealing the unit on loss of air pressure to prevent liquid backflow. A typical tube diffuser assembly is shown in Figure 4.3.

Tube diffusers are designed to operate in the air-flow range of 0.5 to 2 L/s (1 to 4 scfm). Because of their inherent shape, it is sometimes difficult to obtain air discharge around the entire circumference of the tube. In general, less air flow can be expected out of the bottom of the diffuser because the air must discharge against the greatest water pressure. The air distribution pattern will vary with different types of diffusers. In general, however, the extent of inoperative area will be a function of the air-flow rate and the headloss across the media. The dead areas can provide sites for slime growth and other foulant development as discussed later in this chapter.

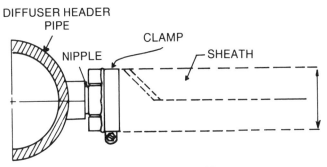

FIGURE 4.3—Typical tube diffuser assembly.

Most tube-diffuser assemblies include a 19 mm (0.75 in.) threaded nipple (stainless steel or plastic) for attachment to the air-piping system. This design makes the tubes especially well-suited for retrofit or upgrade applications because many nonporous diffuser systems use the identical method of attachment. Tube diffusers, however, usually operate at a lower air-flow rate than most nonporous diffusers. Therefore, installation of additional attachment ports on the distribution header are often needed. The depth of tube submergence in the basin will vary. In new installations, the tubes are usually placed as close to the floor as possible, typically within 30 cm (1 ft). In retrofit applications, the discharge pressure of the existing blowers will control the submergence. The tubes will either be installed at the same elevation as the original system or possibly at a somewhat greater distance off the floor to compensate for any increase in headloss through the porous media as opposed to the nonporous device it is replacing.

Historically, tube diffusers have been predominantly installed along one or both long sides of the aeration basin (single- or dual-spiral roll pattern, respectively). It is now becoming common to array these diffusers in a uniform grid pattern to improve OTE. In some cases, the headers are mounted on mechanical lifts. Using this concept, the air headers and diffusers can be removed for inspection and cleaning without dewatering the basin. On the header itself, the tubes can be installed along either one side (i.e., narrow band) or both sides (i.e., wide band) of the pipe. Tubes can also be installed in either a cross-roll or total floor-coverage pattern. In the cross-roll design, the headers are placed across the tank width and the spacing between diffusers, 0.3 to 0.9 m (1 to 3 ft), is small in comparison to the spacing between headers, 3 to 9 m (10 to 30 ft). In the total floor-coverage pattern, the distance between headers and the spacing between diffusers on the headers approach the same value. For relative efficiencies of various diffuser arrangements, see the subsequent section on Tank Geometry and Diffuser Placement.

NONPOROUS DIFFUSERS

Nonporous diffusers differ from porous devices because they use larger orifices or openings that do not easily clog. Nonporous diffusers are available in a wide variety of shapes and materials. The discussion in the subsections to follow addresses fixed orifice diffusers, valved orifice diffusers, static tubes, and serrated hose.

FIXED ORIFICE DIFFUSERS. Fixed orifice diffusers vary from simple holes drilled in piping to specially configured openings in metal or plastic fabrications. Included in this category are perforated piping, spargers, and slotted tubing.

Perforated Piping. Perforated piping is the starting point from which all other orifice devices attempt to improve. These units deliver air through 0.6 to 1.3 cm (0.25 to 0.5 in.) drilled holes spaced at intervals along the piping. Blowoff legs are provided to limit the maximum pressure and to purge liquid from the air piping. Despite their low-transfer efficiency, perforated pipes are occasionally used as for temporary expediency or for unusual clogging conditions, in which quick-cleaning characteristics are more important than air economy.

Spargers. Spargers were introduced in 1954 to provide nonclog performance with transfer efficiencies greater than perforated pipes. A typical sparger-type device is shown in Figure 4.4.

The devices typically are constructed of molded plastic and are saddle mounted below the air header at a center-to-center spacing of 0.3 to 0.6 m (12 to 24 in.). They include a 1.6 cm (0.03 in.) diameter, 15.2 cm (6 in.) blowoff leg below each cluster of four orifices. The

FIGURE 4.4—Typical sparger-type diffuser.

blowoff leg purges liquid and relieves air at high-flow rates or when the orifices on a sparger plug. When plugging occurs, the air flow exits from the drop pipe and poorer oxygen transfer results. It is difficult to tell from the tank surface appearance that anything has happened. A drop in DO or increased air supply could be indicating factors. Orifice sizes range from 0.3 to 0.8 cm (0.1 to 0.3 in.) in diameter. The normal air-flow range for these units is 4 to 6 L/s (8 to 12 scfm) with an extreme range of 3 to 10 L/s (6 to 20 scfm).[1]

One type of sparger is mounted on individual drop piping. An inverted pyramid constructed of molded plastic deflects the air that is released from vertical slots in the built-in blowoff leg. An orifice is provided in the air-supply piping to control the air-flow rate to individual diffusers. Several orifice sizes are available to adjust the flow rate delivered by each diffuser. This type of diffuser is available in drop pipe diameters of 2.5 and 5 cm. The smaller units have an air-flow range of 3 to 15 L/s (6 to 30 scfm), whereas the larger units operate over an air-flow range of 7 to 30 L/s (15 to 60 scfm).

Slotted Tube. Slotted or perforated tubing is sometimes referred to as a wide-band, coarse bubble diffuser. This tubular diffuser is manufactured from stainless steel, which must be thick enough to withstand welding and the vibration structural conditions imposed during operation. Holes in the upper portion of the air reservoir are the intended path of air release. A slot is provided to purge liquid and relieve air during high air-flow rates if the air ports plug. Individual orifices for each diffuser are not provided; but a "balancing nozzle" is provided at the inlet of individual diffusers. A variety of balancing nozzle diameters are available. The units have a desirable operating range of 5 to 12 L/s (10 to 24 scfm).

VALVED ORIFICE DIFFUSERS. Valved orifice diffusers (Figure 4.5) provide a check valve to prevent backflow when the air is shut off. Several of these devices also allow adjustment of the air flow by changing either the number or the size of orifices through which air is released. Valved orifice diffusers are generally mounted on the crown of distribution headers as opposed to the invert of the header like many fixed orifice devices. Consequently, it is good practice to provide a blowoff to purge the header of liquid despite the presence of the check-valve feature. Numerous types of valved orifice devices are available.

One design consists of a 3.6 cm (1.4 in.) diameter molded plastic body with a threaded bottom air inlet; a ball check valve mounted below the orifices in the body; 12 orifices each about 0.3 cm (0.12 in.) in diameter, and arranged in three vertical rows of four orifices; and an orifice adjustment bolt that threads into the body from the top. The orifice adjustment bolt permits varying the number of orifices exposed. These diffusers are screwed into the top of aeration headers and are typically designed for an air-flow range of 3 to 6 L/s (6 to 12 scfm).

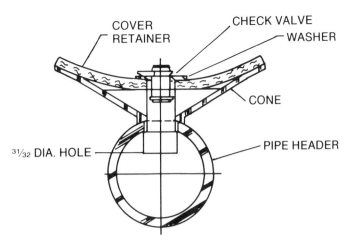

FIGURE 4.5—Typical valved orifice.

The second design consists of a cast body with an inner-air chamber and an inlet-air orifice extending through the threaded bottom connection. The air chamber has a 7.6 cm (3 in.) diameter plastic disk that is retained in position by a stainless steel spring-wire. Air is released around the disk as air pressure in the header lifts the disk from its seat. The disk reseats when the air is shut off. The normal operating air flow for these units is 3 to 6 L/s (6 to 12 scfm).

The third design uses a cone-shaped plastic base and an elastomer cover that flexes around the edge when air is applied, thus letting air escape. As the air flow is reduced or shut off, the cover moves back into place to prevent entry of solids into the diffuser or header. The cover is held in place by a plastic retainer. These units can handle air flows up to 10 L/s (20 scfm).

STATIC TUBES. The static tube aerator is similar to an air-lift pump with interference devices within the tube (Figure 4.6). The intention is to mix the liquid and air, shear the coarse bubbles into smaller bubbles, and increase the contact time between liquid and air. When installed, static tubes are anchored to the basin floor or to a deadman. Air is distributed to the diffusers with submerged piping and released through orifices in the piping at the bottom opening of the tube. As with other devices, the orifices or diffuser distribution devices serve to balance the air flow delivered to each diffuser. Orifices are typically 1.3 cm (0.5 in.) drilled holes. Static tubes may be constructed of polypropylene or polyethylene, and supported with stainless steel anchor legs. The tubes are 0.3 or 0.45 m (12 or 18 in.) in diameter and 0.3 to 1.5 m (1.5 to 5 ft) long. The air-flow rate delivered to each tube depends on the size of the unit. For smaller tubes, air-flow rates of 5 to 10 L/s (10 to 20 scfm) are typical, although air-flow rates of up to 25 L/s (50 scfm) may be used for large tubes.

PERFORATED HOSE. Perforated hose consists of PE tubing that is held on the floor of the basin or lagoon. Slits or holes in the top of the hose release air. The size and spacing of apertures can be varied. The tubes are fed from air manifolds running lengthwise along the side of the basin being aerated. The tubing is installed across the width of the basin.

SYSTEM LAYOUTS FOR NONPOROUS DIFFUSERS. Typical system layouts for fixed orifice and valved orifice diffusers closely parallel the layouts previously described for porous tube diffusers. The most

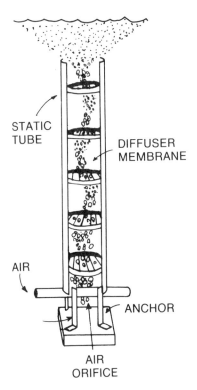

STATIC
TUBE

DIFFUSER
MEMBRANE

AIR

ANCHOR

AIR
ORIFICE

FIGURE 4.6—Typical static tube aerators.

prevalent configurations are the single- and dual-spiral roll patterns using either narrow-band or wide-band diffuser placement. Cross-roll and full floor-coverage patterns may also be used. As with the tube diffusers, mechanical swing-type headers, which allow removal of the diffusers without dewatering, are commonly used.

Fixed and valved orifice diffusers are more commonly employed than porous diffusers in wastewater applications where mixing is more important than oxygen transfer. These include aerated grit basins, channel aeration, sludge and septage storage, and flocculation chambers.

Static tubes are installed in aeration basins or aerated lagoons in a floor-coverage pattern. Typical center-to-center spacing for these units is 1.2 to 3 m (4 to 10 ft).

The most common application of serrated hose has been in shallow-aerated lagoons in which the hoses have typically been installed at center-to-center distances of 1.5 to 6 m (5 to 20 ft). A more recent application is the use of perforated hoses in oxidation ditches. In this approach, the tubes are placed in a denser pattern of 0.15 m (6 in.) center to center.

OTHER DIFFUSED AERATION SYSTEMS

JET AERATION. Jet aeration combines liquid pumping with air diffusion. The pumping system recirculates liquid in the aeration basin, ejecting it through a nozzle assembly. Jet aerators are configured either as cluster aerators or as directional aerators (Figure 4.7). The distribution piping and nozzles are typically of fiberglass construction.

Normally, the recirculation pump is a constant rate device and the power turndown for the aerator is accomplished by varying the air-flow rate.[11] A typical nozzle has a 2.5 cm (1 in.) opening through which the air and mixed liquor pass. Some jet aeration systems are equipped with

AIR IN

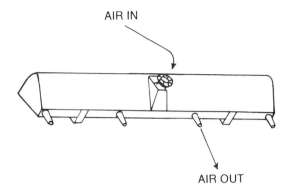

AIR OUT

FIGURE 4.7—Directional jet aeration.

"self-cleaning" features to overcome potential plugging problems. One of these systems uses pneumatic backflush lines and retrievable screens on the pump intake. The backflushing action may be directed toward the distribution pipe to clean nozzles or to the intake screen.

ASPIRATING DEVICES. Another aeration device is the motor-driven, propeller aspirator pump. This device basically consists of a hollow tube with an electric motor on one end and a propeller at the other. The propeller end of the tube is equipped with a guide to direct underwater air flow. The pump draws air from the atmosphere at high velocity and injects it underwater where both velocity and propeller action create turbulence and diffuse the air as bubbles into the water. Pumps can be positioned at various angles depending on basin depth, aeration, and mixing and circulation requirements. The pump is portable and can be mounted on booms or floats in lakes and ponds. Degree of mixing, vector (i.e., initial bubble direction), and speed of aspiration can be controlled. An aspirator pump with a disk rather than a propeller at the end disperses bubbles at a 90° angle to the shaft.[12] Operation of aspirating devices in very cold weather has been reported to be a problem because the aspirator pipe can ice up at the surface, shutting off the air supply.[13]

U-TUBE AERATION. U-tube aeration consists of a shaft from 10 to over 100 m (30 to over 300 ft) deep that is divided into two zones (Figure 4.8). Air is added to the influent wastewater in the downcomer; the mixture travels to the bottom of the tube and then back to the surface where the treated effluent is removed for clarification. The great depths to which the air-water mixture is subjected results in high-transfer efficiencies because the high pressures force all the oxygen into solution. Speece et al.[14] performed a sensitivity evaluation of U-tube aerators that indicated the optimum conditions are a depth of 30 m (100 ft) and a velocity of 2.4 m/s (8 fps).

The cost effectiveness of U-tube aeration relates to waste strength. For normal strength wastewaters (100 to 200 mg/L BOD), what governs the amount of air added is not the oxygen demand, but the amount of air needed to circulate the fluid in the shaft. The air is the motive force for pushing the wastewater around the shaft. For high-strength wastes (i.e., above 500 to 600 mg/L of BOD), however, the amount of air added to the deep shaft no longer depends on the need to push the wastewater around; it is governed by the oxygen demand of the waste. Under these conditions, all the oxygen put into solution is likely to be consumed. Thus, the economics of the deep-shaft method and the power requirements look more attractive as the waste gets stronger.

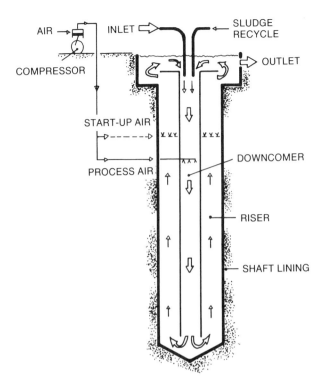

FIGURE 4.8—U-tube aeration.

When evaluating U-tube aeration, a variety of other process considerations must be carefully evaluated in addition to aeration efficiency. These include the reported difficulty in obtaining a well-clarified effluent in some applications.[13] One British application of the process demonstrated that a good effluent quality can be achieved when the U-tube aerator is placed at the inlet end of a plug flow-activated sludge plant where the remaining aeration was provided by a porous diffused air system.[15]

COMBINED DIFFUSED AIR-MECHANICAL AERATORS.
Combined diffused air with mechanical aerators located immediately upstream are used in race track configured aeration basins. The aeration propellers provide liquid circulation and increase turbulence in the diffused-air aeration zone which increases aeration efficiency. Separation of aeration and mixing fuctions in this system enhances process control and ensures optimal treatment unit conditions.

FACTORS AFFECTING DIFFUSED AERATION PERFORMANCE

Diffused aeration performance is affected by many factors including physical characteristics of the diffuser, aeration system configuration, tank geometry, and process considerations. Evaluation of these factors has been the focus of considerable research in recent years. This section briefly summarizes current understanding.

BUBBLE SIZE AND POROUS MEDIA CHARACTERISTICS. In diffused aeration, oxygen transfer occurs during bubble formation, during the bubble's ascent to the surface, and at the water surface itself. Some researchers maintain that most oxygen transfer occurs during bubble formation when the interfacial area exposed to the liquid is constantly being renewed,[16] although others contend that very significant transfer occurs during the bubble's ascent.[17] Regardless of where the transfer occurs and all other things being equal, the rate of transfer is proportional to the area of contact between the liquid and the oxygen. This is the basic advantage of small bubbles. For example, 0.03 cu m (1 cu ft) of air in one bubble would have only 0.46 m^2 (4.8 sq ft) of contact area; but when divided into bubbles of the size commonly produced by porous diffusers, 2 to 3 mm (0.10 to 0.12 in.) diameter, the same volume of air has about 65 m^2 (700 sq ft) of contact area.[1]

There seems, however, to be a limit to the effectiveness of decreasing bubble size. Barnhart found that the overall gas transfer coefficient, K_La, increased while bubble size decreased until the bubble diameter approached 2.2 mm; but, further reduction in bubble size resulted in decreasing K_La.[18] Although smaller bubbles may increase OTE, the additional power required to offset the increased headloss across the diffuser may negate any potential savings.

In general, the size of the bubbles emitted from a clean porous diffuser is controlled by the effective pore diameter, surface tension of the liquid, wetting properties of diffuser surface and gas-flow rate. Of these, the effective pore diameter of the operating pores and the wetting properties of the diffuser material have the greatest impact on bubble size.[19] The wetting property of a diffuser relates to the material's tendency to be hydrophobic or hydrophillic. Materials such as ceramic are hydrophillic and are considered to have good wetting properties, whereas some porous plastic materials tend to be hydrophobic with poor wetting properties. The effective pore diameter of a hydrophobic material must be smaller than a hydrophillic material to produce an equal size bubble.

"Permeability" is one parameter that is commonly specified in attempts to typify pore characteristics of diffuser materials. Permeability is a measure of a porous medium's frictional resistance to flow; it is an empirical rating that relates flux rate to pressure loss and pore size or pore volume.[7] When applied to air diffusers, the term "specific permeability" is typically used. This is determined by a measured air-flow rate that is applied to a diffuser, which is mounted in a manner similar to the way the diffuser would be used at a pressure differential of 5 cm (2 in.) water gauge. From this measurement and the geometry of the diffusers, estimates are then made as to what the air flow (scfm) would have been at 5 cm (2 in.) water gauge differential had the dimensions of the test diffuser been 30 cm × 30 cm × 2.5 cm (12 in. × 12 in. × 1 in.). The dome and disc diffusers that are currently the most popular have specific permeabilities in the range of 258 to 387 L/s-cm/m^2 (20 to 30 scfm-in./sq ft. Unfortunately, permeability measurement does not provide a true basis for comparison of media performance because the same permeability rating could be obtained from a diffuser with a few relatively large pores or a multitude of small pores.[1] The effects of and allowances for other influential factors (e.g., thickness, seal configuration, humidity, temperature, and absolute pressure) are neither known nor considered. Therefore, neither permeability nor specific permeability provides an accurate descriptive characteristic.

Recently, more appropriate test parameters have been developed to measure a diffuser's operative pressure and its effective pore diameter. These are the dynamic wet pressure (DWP) and the bubble release vacuum (BRV) measurements; both parameters are measured when the diffuser is in the wetted state. The DWP parameter is defined as the operating headloss across diffuser media submerged in water at a specified air-flow rate per diffuser.[5] As a general rule, the smaller the bubble size, the higher the DWP. The porous media currently in use have a DWP of 5 to 36 cm (2 to 14 in.) of water when operated within typical or specified air-flow ranges. The specific value depends on the air-flow rate, type of material, diffuser thickness, and surface properties. For the ceramic and plastic materials, the majority of the DWP is associated with the pressure required to form bubbles against the force of surface tension. Only a small fraction of the DWP is required to overcome frictional resistance. Thus, the thickness of the material is only a minor contributor to DWP.

The BRV, as indicated by the name, is a measure of the vacuum in inches of the water gauge required to emit bubbles from a localized point on the surface of a thoroughly wetted porous diffuser element. This is accomplished by first applying a vacuum to a small area on the working surface of a wetted diffuser, and then measuring the differential pressure when bubbles are released from the diffuser at the specified flux rate at the point in question.[20] This test has been applied to a variety of fine pore diffuser elements. It is a sensitive indicator of the fouling degree of porous diffusers and it provides a quantitative means of assessing the rate and location of plugging.

UNIFORMITY. Uniformity of air distribution, both across an entire aeration system and within individual diffusers, is of extreme importance if high OTE is to be attained. On a system-wide basis, uniform air distribution can be achieved through proper sizing of individual control orifices; the orifices balance flow among the diffusers. Air distribution piping and orifice selection must be carefully designed to avoid significant headlosses that may result in unbalanced air flow.

The uniformity of air distribution in an installation may be measured by the timed collection of air in a tank in which the water level is maintained a few inches above the diffuser elevation. The collection vessel is filled with water and inverted over the diffuser to collect the diffused air. The vessel may be of such cross-section to collect either all of the air emitting from the diffuser in question or a known fraction of it. The coefficient of variation of a number of measurements provides a useful comparative parameter.

Concern over uniformity on an individual diffuser basis relates to fine pore diffusers, but particularly to porous media diffusers. The diffuser must be capable of delivering uniform air distribution across the entire surface of its media. If dead spots exist, chemical or biological foulants may form and eventually may lead to premature fouling of the diffuser. Also, if small areas of extremely high air-flux rate are present, larger bubbles may form and OTE will decrease. A porous-diffuser specification should include a requirement for testing to assure that the media will distribute air uniformly.

One is the effective flux rate, which is the weighted average flux rate divided by the average flux rate. Flux rates are determined by timed measurements of collection rate by an inverted vessel of known cross-sectional area. Yet another is the coefficient of variation of BRV measurements at points selected to be representative on an area basis. Another is the coefficient of variation of the values thus obtained.

For tubular diffusers, uniformity may be measured by determining coefficient of variation of measured collection rates at various points down the length of the diffuser, with a sample collector of suitable size and shape. All of these methods are suitable for use either with field installed diffusers or diffusers in the laboratory.

DIFFUSER SIZE AND SHAPE. The results of several studies indicate that there is a correlation between the size or shape of a porous media diffuser and specific oxygen transfer per unit.[6,21,22] These results, however, also may be influenced by the relative uniformity of the diffusers and by how well the various diffusers make effective use of available surface area. The comparisons were made using diffusers constructed of similar ceramic media. Applying the results to diffusers constructed of differing media might not be appropriate.

Data from clean water tests suggest that large diameter ceramic disc diffusers transfer more oxygen than small diameter ceramic domes for a given air-flow rate per diffuser. Yunt compared performance of a 22 cm (8.7 in.) diameter disc diffuser with that of a 17.8 cm (7 in.) dome diffuser.[22] The tests were conducted under similar operating conditions and the diffusers had similar permeability ratings. The results indicate that fewer of the large disc diffusers are required to achieve equivalent oxygen transfer results. The relative number of diffusers seems to depend somewhat on air flow per diffuser. At a rate between 0.5 and 1 L/s (1 and 2 scfm), however, Yunt concluded that the appropriate relative number of disc to dome diffusers would be 0.75 to 0.83 (average 0.79).

Huibregtse, *et al.* also evaluated the effect of diffuser size by comparing the performance of two disc diffusers with effective diameters of 25.4 cm and 21.0 cm (10 and 8.25 in.).[21] The results indicated that increasing the available diffusion area increased transfer efficiency by 3 to 12% depending on diffuser depth. The increase in transfer efficiency was not directly proportional to the increase in diffuser size.

Some tube diffusers make relatively inefficient use of their surface area because only 50 to 60% of the surface area (the upper portion of the tube) is effectively used to transfer oxygen into water.

AIR FLOW PER DIFFUSER. Figure 4.9 demonstrates the affect of air-flow rate per diffuser on SOTE. The transfer efficiency for all types of diffusers are relatively insensitive to air rate; however, all the fine pore diffusers exhibit a measurable reduction in SOTE with increased air rate. The nonporous diffusers are even less sensitive, and exhibit a slight increase in efficiency with increasing air flow. In contrast to

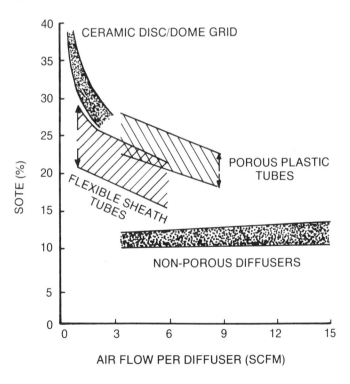

FIGURE 4.9—Effect of air-flow rate per diffuser on SOTE for four diffuser types. Note: Higher SOTE values for one diffuser type at any given flow rate indicates increased diffuser density or dual placement; diffuser submergence is 4.5 m (15 ft).

porous media diffusers, SOTEs for nonporous diffuser systems are relatively unaffected by air-flow rate with indication of increasing SOTE at the higher air-flow rates. The trends shown in Figure 4.9 are very similar to results reported from earlier (1964) tests conducted on spargers and fine pore tubes.[16]

DIFFUSER DEPTH. The effects of depth of submergence for several types of diffused aeration systems are illustrated in Figures 4.10 and 4.11.[23] Although these data are for one specific test tank and air-flow rate, they are representative of the typical effects of depth on performance over the range of depths shown.

In general, the SOTE (Figure 4.10) for diffused air systems increases with diffuser depth because of increased oxygen partial pressure and increased contact time between the bubble and mixed liquor. The overall standard aeration efficiency (SAE) (Figure 4.11), however, may not reflect this improved performance because the power requirements needed to drive the same volume of air through the diffusers at the greater depths will increase. For fine pore diffusers, increasing the liquid depth seems to have no significant effect on SAE. In contrast, nonporous diffusers and jet aerators exhibit a gradually increasing SAE with increasing depth, although not reaching the overall efficiencies demonstrated by the fine pore systems.

Selection of the most economical depth for aeration design must take into consideration several other factors besides aeration efficiency (e.g., available area, land costs, and the difficulty and cost of construction).

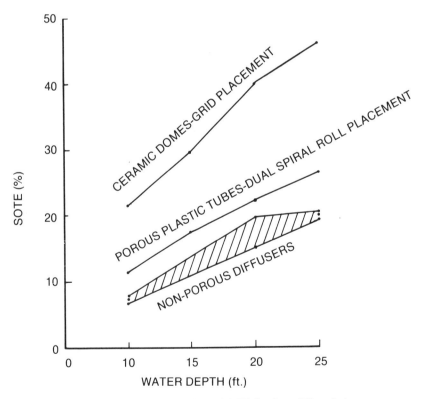

FIGURE 4.10—Effect of water depth on SOTE for four diffused air systems. Note: Tank size was 6 × 6 m (20 × 20 ft); power was 26 W/m³ delivered for tubes and coarse bubble diffusers, and 13 W/m³ for ceramic domes and jet aerators (1 hp/1 000 cu ft and 0.5 hp/1 000 cu ft, respectively).

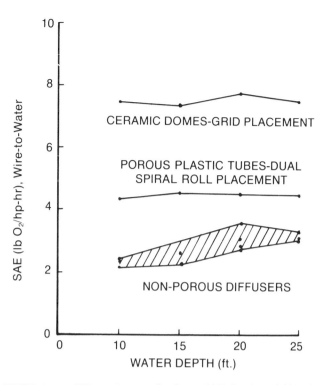

FIGURE 4.11—Effect of water depth on SAE for four diffused air systems. Note: Tank size was 6 × 6 m (20 × 20 ft); power was 26 W/m³ delivered for tubes and coarse bubble diffusers, and 13 W/m³ for ceramic domes and jet aerators (1 hp/1 000 cu ft and 0.5 hp/1 000 cu ft, respectively).

DIFFUSER DENSITY. Diffuser density is defined as the number of diffusers per unit of horizontal floor area. Several researchers have found that increasing the diffuser density for porous media devices at the same air-flow rate per diffuser results in an increase in OTE.[21,22,24,25] This relationship is illustrated in Figure 4.12.[21] When designing new diffuser systems, diffuser density should be maximized within the constraints of minimum allowable air flows and economic costs.[6] Practically speaking, however, some spacing between diffusers is needed to allow access for cleaning and maintenance. Diffuser density should be tapered off in plug-flow tanks concomitant with decreasing oxygen demand to avoid overaeration,[12] unless air requirements for mixing govern.

TANK GEOMETRY AND DIFFUSER PLACEMENT. With a plug-flow aeration system or other processes of a similar nature, oxygen demand varies greatly. At points of high-organic concentrations, it is difficult to maintain a residual DO concentration. High-organic loading may encourage sliming of porous diffusers. Plug flow exacerbates such tendencies because of the localized high-organic loadings experienced in the first pass. Also, in situations in which there are long narrow tanks in multiple-pass series, oxygen demand is lowered to a point at which it is virtually impossible to decrease diffuser density adequately to prevent overaeration and still maintain sufficient mixing.[6] Step feeding has been found to alleviate partially these problems in multiple-pass basins, although data from at least one operating plant indicate that step feeding is less efficient than plug flow in terms of overall oxygen transfer.[26]

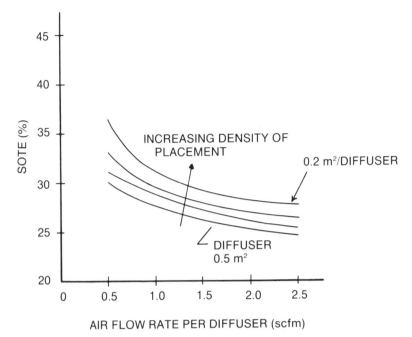

FIGURE 4.12—Effect of diffuser density on SOTE for ceramic disc dome grid configurations. Note: Density of placement varies from 0.2 to 0.5 m²/diffuser (2.2 to 5.4 sq ft/ diffuser); water depth is 4.5 m (15 ft).

Some researchers have recommended consideration of completely mixed tanks for aeration systems using porous diffusers to avoid localized high-organic loadings.[6] If plug flow is used, they recommend design of single-pass tanks with relatively low length to width ratios. Other researchers, however, feel the concern about reduced aeration efficiency because of diffuser fouling in plug-flow tanks may be overstated,[19] and that the process advantages of plug-flow operation (particularly the avoidance of sludge bulking[27]), may outweigh reduced OTE in portions of the basins. In any event, aeration efficiency is only one consideration in selecting process types and basin configurations.

Diffuser placement can also have a significant impact on the SAE of a diffused aeration system. For diffused aeration systems using tubes, socks, or some nonporous diffusers, Yunt found that better aeration efficiency was achieved by means of total floor coverage as opposed to a double-roll system.[17] When roll systems are used, dual-header systems provide better oxygen transfer performance than single-header systems.[21] Results from a similar comparison of diffuser placements for flexible membrane diffusers are shown in Table 4.1.[28] This study also found full-floor coverage the most efficient configuration; the increase in efficiency for quarter point vs. single-spiral roll placement in a rectangular basin was greater than for the mid-width placement. In quarter-point agitation, two rows of diffusers are placed longitudinally in a long rectangular basin, one-quarter the distance from the basin walls.

Similar results have also been found for nonporous diffusers. Rooney et al. compared various diffuser arrangements and found both cross-roll and center-axial arrangements offer significant improvement in OTE over the conventional side-roll arrangement; the cross-roll arrangement was the most efficient.[29] Rooney further found that the effect of diffuser placement increased at greater depths and greater air-flow rates. Schmit et al. also found that mid-width diffuser placement is superior to a single-side–roll placement and that, from an oxygen transfer standpoint,

Table 4.1. Effect of Diffuser Placement on Clean Water Oxygen Transfer Efficiencies (OTEs) of Flexible Sheath Tubes.

Placement	Air Flow (scfm/diffuser)	SOTE (%) at Water Depth		
		3 m	4.5 m	6 m
Floor Cover (Grid)	1–4	14–18	21–27	29–35
Quarter Points	2–6	13–15	18–22	24–29
Mid-Width	2–6	9–11	15–18	23–17
Single-Spiral Roll	2–6	7–11	14–18	21–28

Note: cfm × 0.47 = L/s

deep, narrow basins are preferable.[30] These findings can be attributed to the difference in the mixing pattern achieved. The "column aeration" pattern established by the cross-roll placement and the interference of the spiral-roll pattern caused by mid-width diffuser placement or narrow aeration basins cause a stilling effect that seems to promote oxygen transfer. Many of the mixing patterns and diffuser placements mentioned in the previous discussion are illustrated later in this chapter.

ALPHA SENSITIVITY. Diffused aeration systems are affected by the presence and concentration of certain wastewater constituents such as surfactants.[6] Indeed, failure may result because of inaccurate estimation of oxygen transfer. Wastewater constituents may affect porous diffuser OTE to a greater extent than they do other oxygen transfer devices, resulting in lower alpha factors.[7] One study of a ceramic dome system operating is an essentially plug-flow aeration tank found variations in alpha from 0.3 at the start of treatment (i.e., when the wastewater first comes into contact with recycled sludge) to 0.8 at completion of treatment (i.e., when a fully nitrified effluent is produced).[6] A second study of flexible membrane diffusers, installed in a plug-flow aeration system that was not nitrifying, found that the apparent alpha increased from 0.27 at the aeration tank inlet to 0.42 in the effluent pass.[31] Similar results were found at Whittier Narrows, California, during a comparison of a ceramic grid system to that of a jet aeration system. In the jet system the jets were installed on one side of the basin along the entire tank length with the nozzles being directed across the basin floor in a reverse spiral roll.[32] Figure 4.13[32] is a plot of apparent alpha for each

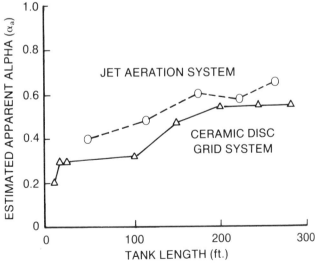

FIGURE 4.13—Estimated change in apparent alpha with tank length, Whittier Narrows, CA.

system versus position from the inlet end. Both systems exhibited similar trends.

DO CONTROL. One of the principal factors affecting aeration efficiency is whether the mixed liquor dissolved oxygen concentration is effectively controlled. This topic is covered in detail in Chapter 6. A typical relationship between mixed liquor DO and aeration efficiency is shown on Figure 4.14. This relationship indicates that close control over DO levels is very important in any diffused aeration system.

FOULING OF POROUS DIFFUSERS

Air Side Fouling. Causes of air-side fouling in porous diffusers include:[1]

- dust and dirt from unfiltered or inadequately filtered air,
- oil from compressors or viscous air filters,
- rust and scale from air pipe corrosion,
- oxidation and subsequent flaking of bituminous air main coatings,
- construction debris, and
- wastewater solids entering through diffuser or piping leaks.

Problems associated with air-side diffuser fouling are considered to be rare; when they are encountered, they can generally be corrected with proper selection and maintenance of air filtration equipment, appropriate air main materials, post-construction cleaning of the process air-handling system, and minimizing blower downtime. These considerations are discussed later in this chapter.

Liquid Side Fouling. All types of diffusers occasionally become fouled with use. Fouling is site and waste specific and is often difficult to forecast.[5] One of the major operational problems is the formation of biological slime on the external surface of the diffuser. Although it is difficult to draw generalizations, several researchers have identified potential causes of biofouling, including:[5]

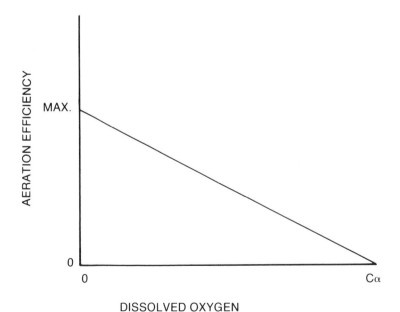

FIGURE 4.14—Relationship between aeration efficiency and dissolved oxygen concentration.

- high organic loading,[6,33]
- low DO concentration,[6]
- waste characteristics,[6]
- high waste temperatures,[33,34]
- low air flow rate per diffuser,[33]
- uneven air distribution in diffuser,[2,33,34]
- low permeability of diffusers,[33,36]
- presence of inorganic particulates,[7]
- nutrient imbalance, and
- chemical additives such as iron or effluents from demineralizations.[7]

Houck and Boon speculated that bioslime formation primarily results from high-organic load or low DO and that the growth is exacerbated by extreme plug-flow aeration tank design.[6] The latter point is supported by the work of Addison in Ontario.[37] Low air-flow rates through the diffuser also seem to have a significant detrimental effect when other fouling conditions are present. Most manufacturers of disc and dome diffusers recommend maintaining a minimum flow rate of 0.25 to 0.35 L/s (0.5 to 0.7 scfm) and minimizing interruption of the air supply.

The second major concern is external plugging because of chemical precipitates, such as calcium carbonate ($CaCO_3$). This phenomenon seems to have occurred at the Central Contra Costa plant in California,[38] the Basingstoke, UK plant,[6] and Whittier Narrows, California.[39] Plugging results from shifting the bicarbonate equilibrium and precipitation of $CaCO_3$. This should be anticipated in hard alkaline waters. Plants with recycle streams containing ferric chloride ($FeCl_3$) may also experience scaling problems caused by precipitation of iron salts. Usable correlations between scaling tendencies and indicators such as alkalinity, Langlier index, and Ryzer index have not been found.[19] The scaling tendencies of porous diffusers, however, may be quantified by tests on pilot diffusers, which may be placed in an operating plant and which are removable without interrupting the process. As an alternative, diffusers removed after draining may also be evaluated. The presence of a scaling tendency does not preclude use of porous diffusers because effective-cleaning techniques are available, as described in Chapter 7. It does increase the maintenance requirements, however, and this cost factor should be carefully evaluated when considering the relative economics of alternative aeration systems.

EFFECTS OF DIFFUSER FOULING. Diffuser fouling is generally detrimental in wastewater treatment. It may result in increased headloss across the diffuser and reduced OTE, both of which result in increased power consumption to maintain a given level of treatment. The results of the current EPA-ASCE Research Project suggest that the incidence and consequences of fouling of fine pore diffusers is significantly less severe than has previously been supposed.[40]

Operating experience at several plants indicates that, despite the presence of biofouling, porous disc and dome diffusers do not exhibit significant increases in headloss.[9] These same plants, however, do exhibit dramatic reductions in OTE, as great as 50% because diffuser sliming manifests itself as coarse bubbling. One possible explanation is that the presence of bioslime causes the bubbles to coalesce during formation. It is considered more likely that the larger bubbles result from the higher localized flux rates that occur as the diffusers foul. Another theory is that the slime gradually blocks off air flow passages in the media and forces air to short circuit past gaskets or around bolts.[5] The high flow around gaskets and bolts is primarily caused or augmented by changes in dimensions of gaskets or bolts that occur with time in service. However, at most of the sites in the current ASCE EPA Study, diffuser fouling did not result in undue losses in efficiency and could be effectively managed by periodic maintenance procedures.

Although most data indicate sliming produces little change in headloss, there are incidences at sites in Madison, Wisconsin, Beddington, United Kingdom (UK) and Whittier Narrows, California, where severe fouling produced dramatic changes in headloss. At Madison, heavy biofouling of a dome diffuser increased the headloss across the ceramic media from 15 cm (6 in.) to about 46 cm (18 in.) and dramatically altered the air-flow profile.[5] Once a clogging mat had developed within the main body of the dome, the majority of the air passed out along the periphery of the center bolt and along the edges of the dome.

At Whittier Narrows, disc diffusers operated for nearly 3 years without significant increases in headloss (approximately 0.1 psi/yr). After the third year, however, headloss increased dramatically, finally reaching 117 cm (46 in.) of water column at the design flow rate of 0.63 L/s (1.25 scfm). Clean diffuser headloss was 25 to 28 cm (10 to 11 in.) at the same air-flow rate. The cause of the headloss was attributed to the formation of a hard, white substance containing $CaCO_3$, aluminum and grease that adhered to the diffusers.[39]

FOULING AND CLOGGING OF OTHER SYSTEMS. Nonporous diffusers are used in aeration because maintenance is minimal. Clogging from the air side generally is not a consideration, although there have been reports of air distribution piping becoming filled with dried mixed liquor solids.[36] Clogging from the liquid side is infrequent; thus, air headers for nonporous diffusers are commonly fixed to the floor of the aeration basin. Aeration basins have a tendency to coagulate inert debris into large bundles or rope-like groupings. Therefore, there is concern, especially with the static tube devices, that the aerator may collect the debris and ultimately cause problems.

Some clogging problems have been reported for jet aerators. In these instances, the nozzles became plugged with rags or other coagulated inert material. As previously described, some designs include features that reportedly alleviate this problem.

DIFFUSER CLEANING. Various diffuser-cleaning methods have been proposed in the past 50 years. These alternative methods fall into three broad categories:[7]

- methods requiring diffuser removal,
- *in-situ* methods that interrupt process, and
- *in-situ* methods that do not interrupt process.

The effectiveness of these methods, along with their advantages and disadvantages, are discussed in Chapter 7. Regardless of the methods chosen, aeration tanks employing fixed diffuser systems should include means for rapid draining.

MATERIAL BREAKAGE. Diffused aeration equipment must be durable. Material failures can threaten system reliability, increase maintenance costs, and seriously reduce OTE.

Over the years, a variety of material failure problems has been reported for diffused aeration equipment. Many of these relate to the use of PVC, a relatively fragile, temperature-sensitive material. Examples include cracked pipes and pipe saddles, and broken pipe straps and blowoff lines. Problems can be avoided by careful design and installation of the systems, as discussed later in this chapter. Also, consideration should be given to the use of stainless steel appurtenances.[41] Specifying 316 stainless steel for items such as anchorage straps and bolts will add little to overall system cost, but may significantly increase the mechanical and structural integrity of the system.

Significant problems for dome-type diffusers have been breakage or stretching of the retainer bolt. Bolt breakage can be minimized during installation by using a torque wrench that will spin free at preset torque well below the failure point of the bolt. Bolt stretching tends to occur to some extent with all nonmetal center bolts and can result in air leaking past the diffuser gasket. This can be corrected by tightening the bolts down after a period. However, this requires draining the tanks. Center bolts constructed of metal such as brass or stainless steel are available and they can be used to eliminate bolt stretching. Care must be taken to avoid overtorquing the bolts and cracking the diffuser stone, PVC holder, or PVC pipe.

Other concerns with material failure include:
- fatigue of flexible components (e.g., the elastomeric cover);
- ultraviolet deterioration of sensitive plastics;
- structural failure at threaded diffuser connections to air headers;
- creep or compression set in plastic materials; and
- guides and anchor damage because of inadequate allowance for thermal expansion and contraction.

For alternative systems with similar OTEs, equipment without moving parts that are subject to fatigue offer potential advantages from a maintenance and operational standpoint.

PERFORMANCE OF DIFFUSED AERATION SYSTEMS

OXYGEN TRANSFER EFFICIENCY IN CLEAN WATER. Typical clean water SOTEs for a variety of diffused aeration systems are shown in Table 4.2.[7] The data were derived from eight different clean water studies and are reported for a common diffuser submergence of 4.6-m (15 ft).[7] Some but not all of the data were generated using the current ASCE recommended clean water standard.[42] Understanding the accuracy and limitations of the testing method used to generate the data is very important when reviewing transfer data from either clean water or process water conditions. The results of one approach are

Table 4.2. Clean Water OTE Comparison for Selected Diffusers.

Diffuser Type and Placement	Air-Flow Rate (scfm/diffuser)	SOTE (%) at 15-ft Submergence	Reference
Ceramic Discs-Grid	0.6–3.0	30–34	21
Ceramic Domes-Grid	0.5–2.5	25–37	21–23,25
Porous Plastic Tubes			
Grid	2.4–4.0	28–32	46
Dual-spiral roll	3.0–9.7	18–28	21,23,47
Single-spiral roll	2.1–12.0	13–25	21,47
Flexible Sheath Tubes			
Grid	1–4	22–29	28
Quarter points	2–6	19–24	28
Single-spiral roll	2–6	15–19	28
Jet Aeration			
Side Header	54–300	15–24*	23
Flexible Sheath			
Disc, 9 in. Grid	1–6	n.a.	13,26–30,33–39
Disc, 13 in. Grid	4–12	n.a.	21–24,53
Nonporous Diffusers			
Dual-spiral roll	3.3–9.9	12–13	23,48
Mid-width	4.2–45	10–13	23,48
Single-spiral roll	10–35	9–12	23,48

*Energy requirement for pumping not included in SOTE calculation.
Note: ft × 0.3 = m; cfm × 0.47 = L/s

often not comparable to those derived using a different approach. (See, for example, the nonsteady state results reported by Mueller.)[43]

The data in Table 4.2 clearly show the effects of several of the factors previously discussed such as diffuser type and placement. In general, ceramic domes and discs were found to demonstrate slightly higher clean water transfer efficiencies than typical porous plastic tubes or flexible sheath tubes in a grid placement. SOTEs for tubes, discs, and domes are significantly superior to those for all placements of nonporous diffusers discussed in this chapter. A clean water study of jet aeration systems found that under similar operating conditions (i.e., diffuser depth and delivered power level per basin volume), the SOTE for jet aeration was higher than that for nonporous diffusers but was less than the porous media results.[23] Jet aeration systems, however, have an energy requirement associated with pumping. When this is taken into consideration, clean water testing has found the SAE for jet aeration to be similar to that for nonporous diffusers.[23] These relationships were shown earlier in Figures 4.10 and 4.11.

OXYGEN TRANSFER EFFICIENCY IN PROCESS WATER.

Data for process water performance of diffused aeration systems are much more limited than that for clean water performance but are rapidly becoming more available. Unfortunately, this is the information truly of interest in designing and operating aeration systems.

Before 1981, the methods used to evaluate aerator performance under process conditions were inconsistent; coherent data on process water performance were extremely limited.[7] Since that time, numerous studies have been undertaken by the ASCE Oxygen Transfer Standards Committee to identify a more accurate and useful testing method. This effort has led to the development of the off-gas analysis procedure,[44] which is capable of measuring localized performance throughout the basin with respect to OTE and air-flow rate. The technique is well-suited for evaluating the process water performance of diffused aeration systems in plug flow as well as in complete mix tanks.

Table 4.3, presents process water oxygen transfer data from 13 evaluations at various sites employing a variety of aeration systems.[7] The oxygen transfer data were all collected using the off-gas test procedure.[42] Apparent values of alpha were estimated from clean water performance data for similar tank geometry, air-flow rate per diffuser and diffuser placement. Because the performance of most porous diffusers is likely to change with time, the term *apparent alpha*, $\propto_a$, has been adopted to distinguish between differences in clean and process water performance for specific cases in which the diffusers are process tested at a condition of undetermined fouling ($\propto_a$) versus, those in which they are process tested new or just after cleaning ($\propto$). The later condition measures the alpha value because of wastewater characteristics only. In all cases, the OTE (field results) values have been converted to $\propto_a$ SOTE values (i.e., to 20° C), and zero residual DO.

The data presented in Table 4.3 indicate that the differences between apparent alpha values for ceramic grid, fine pore diffusers and apparent alpha values for other more turbulent systems (e.g., jet aerators and static aerators) may not be as great as previously reported in the literature.[45] In addition, the overall average apparent alpha values presented in Table 4.3 and elsewhere[46-52] are lower than many alpha values historically used for design purposes.

When reviewing the data presented in Table 4.3 it is important to realize that the intent of presenting these data is to give the reader a general feeling for the range in performance of the systems listed under a variety of operating conditions.[7] No attempt has been made here to correlate oxygen transfer performance to process type, process loading, wastewater characteristics, and other factors. Also, each data set represents the observed performance of a particular system over a period of several hours only and is not suitable for the design of similar systems.

Table 4.3. Process Water OTE Comparison For Selected Aeration Systems

Site	System	Flow Regime	α_a SOTE (%) Mean	α_a SOTE (%) Range	Diffuser Submergence (ft)	Variation in α_a SOTE	Estimated α_a Mean	Estimated α_a Range	Mean Air-Flux Rate (scfm/sq ft)
Madison, WI	Ceramic grid	Step feed	17.8	12.6–26.2	14.8	Rising from inlet to outlet	0.64	0.42–0.98	0.28
Madison, WI	Ceramic and plastic tubes	Step feed	11.0	7.5–13.4	15.0	Rising from inlet to outlet	0.62	0.46–0.85	0.53
Madison, WI	Wide-band, fixed-orifice nonporous diffusers	Step feed	10.0	7.9–10.9	15.0	Random	1.07	0.83–1.19	0.53
Whittier Narrows, CA	Ceramic grid	Plug flow	11.2	9.3–15.2	13.5	Rising from inlet to outlet	0.45	0.35–0.60	0.21
Whittier Narrows, CA	Jet aerators	Plug flow	9.4	7.8–10.8	13.5	Rising from inlet to outlet	0.58	0.48–0.72	0.37
Brandon, WI	Jet aerators	Complete mix	10.9	9.7–12.1	12.5	Random	0.45	0.40–0.50	0.13
Brandon, WI	Jet aerators	Complete mix	7.5	7.4–7.7	12.5	Random	0.47	0.46–0.48	0.39
Orlando, FL	Wide-band, fixed-orifice nonporous diffusers	Complete mix	7.6	6.8–8.4	13.0	Random	0.75	0.67–0.83	0.92
Seymour, WI	Ceramic grid	Plug flow	16.5	12.0–18.8	13.8	Random	0.66	0.49–0.75	0.07
Lakewood, OH	Ceramic grid	Plug flow	14.5	12.4–15.9	13.3	Rising from inlet to outlet	0.512	0.44–0.57	0.14
Lakewood, OH	Ceramic grid	Plug flow	8.9	7.0–11.1	13.3	Rising from inlet to outlet	0.31	0.26–0.37	0.09
Brewery	Ceramic grid	Complete mix	14.2	12.5–15.2	19.2	Uniform	0.37	0.32–0.37	0.30
Brewery	Static aerators	Complete mix	7.4	5.7–8.8	19.8	Uniform	0.50	0.36–0.51	0.53
Nassau Co., NY	Flexible membrane tubes	Plug flow	7.6	5.7–8.7	13	Rising from inlet to outlet	0.36	0.27–0.42	0.55

NOTE: ft × 0.3 = m; cfm/sq ft × 305 = L/m² • min

For more specific information on the operating conditions associated with the reported test results, the reader is referred to the references cited.

HEADLOSS AND AIR FLOW RANGE. A summary of typical air-flow ranges for a variety of diffuser types is presented in Table 4.4. For uniform air flow to the diffusers, individual orifices are normally used. This orifice usually produces most of the headloss associated with a diffuser element; it should be sized to produce a large headloss in comparison to losses in the air supply system. Typically, these orifices are sized from 0.3 to 1.9 cm (0.12 to 0.75 in.) in diameter. One manufacturer of a nonporous diffuser provides the relationship between headloss, air-flow, and orifice size shown in Figure 4.15. A typical headloss curve for a dome diffuser is shown in Figure 4.16. The slope of the headloss versus air-flow rate curve for the media alone is very flat, thus, necessitating an individual control orifice.

Some porous diffusers (e.g., ceramic domes and discs) have a minimum air flow below which they should not be operated and a maximum air flow above which aeration efficiency decreases more rapidly. For dome and disc diffusers, the minimum flow rate is 0.25 to 0.38 L/s (0.50 to 0.75 scfm), depending on diffuser size. At this point, the air flow should be sufficient to uniformly distribute air across the diffuser and minimize surface growth, infiltration of mixed liquor, and potential fouling problems. An air-flow rate of 0.25 L/s (0.5 scfm) is also the approximate rate at which the distribution of air among diffusers on a grid can be achieved using a typical control orifice. Maximum air-flow rates are determined based on the desired transfer efficiency and the headloss across the diffuser element. As the air-flow rate is increased the OTE decreases, although the headloss increases very rapidly because of the small diameter control orifices typically used in disc and dome systems. Many designers of dome diffusers believe that an air flow of 1.0 to 1.3 L/s (2.0 to 2.5 scfm) is an appropriate upper limit for use during peak load periods (for large disc units, higher upper limits are used). Using these guidelines results in an air-flow, turndown ability of 4 or 5 to 1. Aeration efficiency between air flow rates of 0.3 to 1.0 L/s (0.6 and 2.0 scfm), however, has been found to decrease about 14% for dome diffusers.[23] Therefore, actual oxygen transfer turndown ability would be closer to 3.4 or 4 to 1. Interestingly, nonporous diffusers may exhibit the opposite trend, i.e., change in OTE may exceed the ratio of air-flow ranges.

MIXING CHARACTERISTICS. Mixing serves to maintain solids in suspension and to assure that oxygen is well distributed throughout a tank. Some types of aeration devices can distribute sufficient oxygen to

Table 4.4. Typical air-flow ranges for selected diffusers.

Diffuser Type	(scfm/Diffuser)	Headloss* (inches)
Ceramic Dome	0.5 to 2.5	6–25
Ceramic Disc (8.5 in. Dia.)	0.7 to 3.0	5–19
Porous Media Tube	2 to 6	
Flexible Sheath Tube	2 to 6	
Flexible Sheath Disc-Type I (8.5 in. dia.)	2 to 6	9–23
Flexible Sheath Disc-Type II (9 in. dia.)	2 to 6	
Flexible Sheath Disc-Type III (29 in. dia.)	2 to 20	9–24
Sparger	8 to 12	6–9
Slotted Tube	10 to 24	3–13
Valved Orifice	6 to 12	5–12

*Headloss value is for new, clean diffuser
Note: cfm × 0.47 = L/s; in. × 2.54 = cm

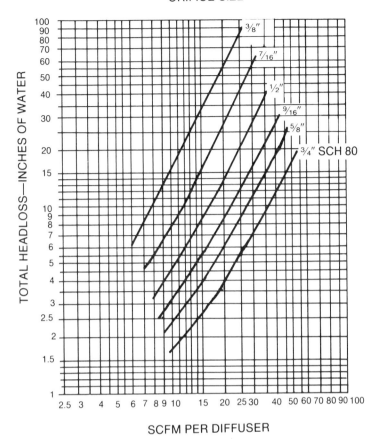

FIGURE 4.15—**Typical headloss characteristics of non-porous diffuser orifices.**

all parts of the tank without providing adequate suspension of solids and without mixing completely. Complete mixing assumes that the influent is spread throughout the aeration tank in a very short time relative to the hydraulic retention time.[49]

Various indicators of mixing efficiency have been proposed (e.g., turnover time, pumping capacity and power per unit volume). These values, however, fall short of a good description of the operation without specification of baffling and other physical features of the system.[49] Mixing is best indicated by tracer studies. Another good indicator in complete mix reactors is the oxygen uptake rate. In an ideally mixed tank, the value would be the same at all points. Suspended solids measurement does not provide a good indication of mixing except in extremely poorly mixed conditions.[49]

Different diffused aeration equipment and diffuser placements produce a variety of mixing patterns. Some of these are illustrated in Figure 4.17. The full-floor coverage configuration mixes in a near vertical pattern with small eddy circulation and with limited interaction between diffuser "cells." Thus, solids are suspended efficiently and there is little bulk mixing. In some retrofit applications using porous diffusers in full-floor coverage, blower limitations have resulted in diffusers being located 0.6 to 0.9 m (2 to 3 ft) off the basin floor. When the aeration tanks are preceded by primary clarifiers, Yunt and others have found no problem with solids deposition below the diffusers.[26]

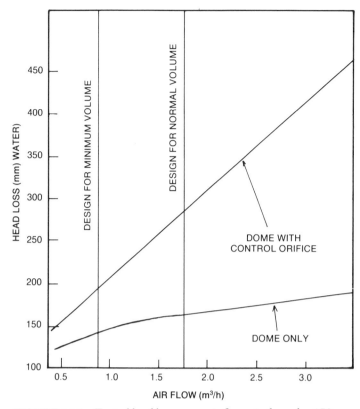

FIGURE 4.16—Typical head loss versus air flow rate through a 178 mm diam dome diffuser.

Spiral-roll patterns provide effective bulk mixing. The mixing characteristics of some systems such as jet aeration or use of a cross-roll pattern are not well documented.

*D*IFFUSED AIR SYSTEM DESIGN CONSIDERATIONS

AIR REQUIREMENTS. Process air requirements and the methods for calculating these needs are outlined in Chapter 2. Once the oxygen demand is estimated and the field OTE of the device is selected, the air requirements may be determined as follows:

$$Q_a = \frac{OTR_f}{.075 \times .232 \times 1440 \times OTE_f}$$

where:
- Q_a = volumetric flow rate in scfm
- OTR_f = field oxygen transfer rate = RV in lb/day
- OTE_f = field oxygen transfer efficiency of the device in %
- $.075$ = approximate density of air in lb/ft^3
- $.232$ = weight percent of oxygen in air

Standard conditions are stated at 14.7 psia, 68°F and 36% relative humidity. Because these conditions are not usually met in the field, conversions of gas flow to field conditions must be made, with corrections for temperature, pressure, and relative humidity. A simplified equation for this conversion is:

$$Q_f = Q_a(.0276)\frac{T_1}{P_1 - (RH_1 \times PV_1)}$$

where:

Q_f	=	the actual volumetric air flow, acfm
T_1	=	the blower inlet temperature in °R
P_1	=	blower inlet pressure in psia
RH_1	=	relative humidity at blower inlet, %, and
PV_1	=	vapor pressure of water at T_1 in psia.

Air Requirements For Mixing. Sufficient air must be provided to prevent deposition of suspended matter. Where wastewater has had all grit removed, a velocity of about 0.15 m/s (0.50 fps) across the tank bottom is adequate. This parameter, however, is difficult or impossible to measure for some aeration systems. In evaluating mixing requirements, different diffuser configurations exhibit very different mixing characteristics. One manufacturer of diffused aeration equipment uses a minimum mixing intensity of 0.17 to 0.25 L/m³ · s (10 to 15 scfm/1000 cu ft) for a grid system and 0.25 to 0.42 L/m³ · s (15 to 25 scfm/1000 cu ft) for a spiral roll (actual value depends on tank width and header arrangement).[5] Another rule of thumb for spiral-roll systems relates air requirements to the length of the tank. Typical design values range from 4.6 to 10.8 L/m · s (3 to 7 scfm/ft).[5]

DESIGN FOR EFFICIENCY AND FLEXIBILITY. The energy efficiency of a system is critically dependent on matching component sizing and flexibility with the process demand. Excess capacity that cannot be adjusted to match the minimum necessary oxygen demand of the process can represent a substantial, unnecessary loss of energy. This consideration is particularly important for relatively small plant sizes (less than 3800 m³/d, or 10 mgd) in which operator attention and component multiplicity are minimal.[55]

Aeration systems are typically designed to meet future loadings that are significantly higher than both initial and average loadings over the life of a facility. Often, facilities are designed to meet peak demands under these future loading constraints. Consequently, the installed aeration capacity can exceed actual demand by ratios of 3- to 6-to-1.[55] Unless considerable flexibility allows closely matching the oxygen supply to the process demand, energy will be wasted. For diffused air systems, operational flexibility can be provided in several ways as described in the following paragraphs.

Multiple Aeration Basins. Using this approach, basins can be placed in service when needed or removed from service when the long-term demand is expected to be low (e.g., seasonal load variations). For systems using PVC distribution piping, special precautions are needed to protect the piping from damage because of expansion, contraction and degradation caused by ultra violet rays from direct sunlight which react with compounds within PVC and make the pipe brittle. Also, for porous media systems, special precautions are needed to prevent deterioration caused by fouling or chemical precipitation.

Optimizing Diffuser Turndown. Diffused aeration can be adjusted over a relatively broad range by controlling air-flow rates. Minimum air-flow rates are typically dictated by mixing requirements or fouling considerations and maximum rates by diffuser transfer efficiency and headloss limitations. The upper and lower limits vary for different devices and system designs. Designing an efficient aeration system requires selecting an air supply and diffuser-operating range that is sufficiently wide to provide efficient operation at low and normal loadings while still meeting air requirements at peak conditions. Consideration should be given to the use of multiple blowers of different sizes to achieve an acceptable operating range. In general, a maximum-to-minimum air-flow ratio of 5:1 is desirable. Reduced aeration efficiency at the higher unit flow rates (a porous diffusion phenomenon) is acceptable during peak conditions because of the infrequency of this occurrence.

Adjustable Number of Diffusers. During initial construction, the number of diffusers installed should be based on anticipated oxygen requirements for the first several years of operation with provisions made for increasing the number of diffusers as loadings increase. Depending on aeration configuration, this can be accomplished by installing diffuser "blanks" or plugs, or by providing for additional, future headers.

Automated Air Flow Adjustment. Automatic controls can be incorporated into diffused air systems to automatically adjust air supply to a preset DO concentration. This topic and other control measures are addressed in Chapter 6.

Staged Construction. Staging construction and building smaller initial facilities allows reduction of the installed capacity to demand ratio.

Performance and operability are two important considerations in the design of flexible aeration systems. Minimum mixing requirements, minimum oxygen supply without affecting effluent quality, and blower-operating range (with consideration given to diffuser plugging) are performance constraints that must be evaluated carefully. System design should balance these considerations with ease of operation. The prime objective of system flexibility is not met if the facilities are not operated to match the supply to demand. If equipment is inherently difficult to take out of service or to place in service or if operating adjustments are difficult, efficient operation will probably not be achieved. [55]

SELECTION OF AIR DIFFUSER TYPE. Selecting the most appropriate diffused aeration system for a given application is a complex and often difficult task. In evaluating alternatives, the following factors should be given careful consideration both from an operational and economic viewpoint.

Oxygen Transfer Efficiency. A system that can reliably transfer oxygen at a high efficiency is obviously desirable. However, the oxygen transfer efficiency of the system must be evaluated on the basis of its performance under actual process conditions (e.g., when the diffusers are dirty and during loading fluctuations). Consideration should also be given to the degree of flexibility and control provided in the aeration system; if air supply is not closely matched to oxygen demand, potential energy savings of one system over another may not be realized.

Maintenance Requirements. In evaluating maintenance requirements, the following questions should be among those answered. Does the system require close monitoring by skilled personnel? What is the frequency of diffuser cleaning or replacement? What is the cost and effort associated with diffuser cleaning and replacement? Is material breakage or fatigue a problem? Does the system require special effort for air cleaning? Will plant personnel have sufficient skill, budget, time, and motivation to fulfill maintenance requirements? Does the aeration system fit in with the overall process design?

Existing Conditions. When installing aeration systems in existing tanks, one must consider the ease of retrofitting, the suitability of basin geometry to various aeration arrangements, and the ability to match mixing and oxygen requirements. Fixed-floor systems require the ability to take a basin out of service without adversely affecting process performance.

The capability of existing air supply systems must also be considered. Often, the diffusers being installed exhibit higher headloss characteristics than those being replaced. Also, an allowance should be made for increased headloss associated with potential diffuser fouling. This may require raising the diffusers up to a few feet above the basin floor or it may necessitate modifying the air supply equipment.

Appurtenances. Any evaluation should also consider the capital and operating cost for any special appurtenances associated with the various aeration alternatives (e.g., air-filtering facilities, basin-dewatering facilities, special piping systems, and special diffuser-cleaning systems).

DIFFUSER PLACEMENT AND MOUNTING. Diffuser placement and mounting can greatly affect aeration efficiency and are an important design consideration.

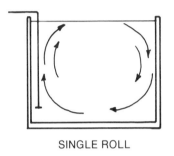
SINGLE ROLL

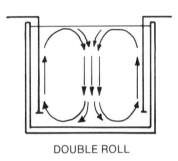
DOUBLE ROLL

SPIRAL ROLL

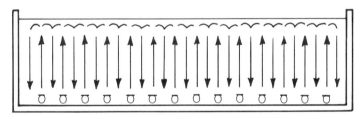
FULL FLOOR COVERAGE

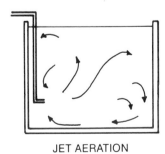
JET AERATION

FIGURE 4.17—Aeration mixing patterns (sections).

Diffuser Placement. The most efficient diffuser placement configurations are full-floor coverage, cross roll, and double roll (air supply either mid-width or along both walls). The least efficient configuration is the side roll (spiral roll)(Figure 4.17).

Placement configurations and the diffusion patterns they create are well documented in manufacturer literature and are dependent to an extent on the basin geometry. The aeration efficiency of some configurations also is influenced by basin geometry. For example, the performance of spiral-roll and double-roll systems can be increased through use of narrower basins. Also, the efficiency of all coarse bubble systems increases using deeper basins.

The selection of the diffuser placement is sometimes influenced by regulatory requirements. Many existing side-roll systems are the result of local requirements that aeration systems must be removable without tank dewatering and without process interruption.

Diffuser Mounting. Diffuser mountings may be either fixed or retractable. Fixed mountings are available in a variety of configurations. Retractable mountings for porous or nonporous media may be obtained in lengths of up to 9.2 m (30 ft). Portable hoists may be used for raising the headers out of the tank for servicing. Because of piping requirements, use of mechanical liftouts is usually impractical or uneconomical for cross-roll and total floor-coverage designs.

Most nonporous diffusers should be mounted on the invert of the air distribution lateral to ensure liquid drainage. If an alternative mounting is used, consideration should be given to providing a bottom-mounted, moisture blowoff line.

The key to successful diffuser mounting is specifying quality hardware and closely supervising installation. Necessary characteristics for diffuser holders are: (a) adequate support to prevent breakage and to guard against blowing out because of differential air pressure; (b) tightness against air leaks that would disturb good air distribution and against mixed liquor infiltration during air interruptions; (c) resistance to disintegration or production of material that could clog diffusers on the air side; and (d) adequate adjustability to allow leveling of all diffusers, thus enhancing good air distribution.

PIPING SYSTEM DESIGN. Piping system design includes air distribution and control as well as piping materials.

Air Distribution and Control. Good air distribution is critical to efficient aeration and effective distribution control is necessary to match air supply to oxygen demand. Considerations made during design and construction can greatly benefit operation and operating economics.

Air-flow metering should be provided at major groupings of diffusers to control air distribution and to monitor the air-flow rate per diffuser or aerator to ensure these devices are operating within a proper air-flow range. Simple indicating flow gauges are often appropriate with throttling valves to adjust flow rate. Air to major process units may justify more sophisticated meters with recording and totalizing devices. A wide variety of flow measurement devices are commercially available. Some common air measurement devices include orifice plates, venturi tubes, flow nozzles, and pitot tubes. References on making flow measurements are L.K. Spink's *Principles and Practices of Flow Meter Engineering*[56] and the gas flow measurement section of the *Proceedings of the Workshop Toward an Oxygen Transfer Standard.*[57]

Pressure gauges at major groupings of diffusers are especially important for porous media diffusers and to a lesser extent provide relevant information for nonporous diffusers. Valves for flow regulation and shut-off should be provided to each group of diffusers.

Sizing of air mains should result in minor friction losses in individual air headers. Many design engineers limit the loss in individual air supply laterals to 1% of the static headloss. In addition to the restrictions on pressure losses in a system, the designer also must recognize the economic consequences of pipe sizing.

Diffusers should be set accurately to the same elevation. Diffusers fed through the same valve should not vary more than 0.6 cm (0.2 in). Also, porous media diffusers fed through the same valve should have approximately the same dynamic wet pressure. Control orifices should be installed on each diffuser to create a significant headloss relative to piping losses; thus the air supply to the diffusers is balanced.

Pipe Materials. When selecting pipe materials and connections, consideration is given to heat, internal and external corrosion, expansion and contraction, and resistance to all normal and abnormal forces.

Blower discharge piping experiences high temperatures. Depending on blower efficiency, the temperature rise above ambient because of air compression is 8 to 12 °F per psi compression. For conventional 5 m (15 ft) deep tanks temperatures of 60 to 80°C (140 to 180°F) are common. In areas of warm climate, however, the combination of high ambient temperature and use of deep aeration tanks can easily raise the blower discharge temperature above 95°C (200°F). Consequently, discharge piping should be insulated or otherwise provided with barriers for personnel protection. The high temperature protects the inside of the piping from condensation; therefore, internal pipe corrosion where the pipe is above water or ground level should be minimal unless unusual conditions exist such as a very moist air supply or long pipe runs.

External surfaces of air piping above grade should be either painted steel, stainless steel, or cast iron. Paint should be suitable for the highest discharge temperature anticipated. Internal surfaces of air piping above the liquid level are often left bare. Corrosion scale that may develop before the pipe is placed in service should be removed. Sometimes a temporary grease coating is used on the pipe interior that should be removed immediately before placing the pipe into service. For porous diffuser systems, scaling and flaking can result in premature diffuser plugging.

Piping in the aeration basin, especially in the vertical drop pipe that penetrates that liquid surface, is subject to corrosion and to wet, oxidized conditions both internally and externally. Corrosion is severe and galvanizing and coal-tar paints provide only short-term protection. Pipe materials are often stainless steel, fiberglass, or plastics suitable for higher temperatures. In some cases, performing heat loss calculations ensures that the air temperature is within the material's specifications. The compounds of PVC that should be used are described in ASTM D-3915, cell classifications and 124524.[7] The PVC should be UV stabilized with 2% minimum TiO_2 or equivalent. Because of its sensitivity to sunlight, PVC should only be used below normal water levels. The specifications, dimensions, and properties of the pipe itself should conform to either ASTM D-2241 or D-3034. Other materials used include mild steel or cast iron with external exotic coatings (e.g, coal-tar epoxy or vinyl). Interior surfaces include cement coatings or exotic coatings. Piping is held in place with stainless steel anchors and bolts.

Because the piping is subjected to wide temperature differences, pipe expansion must be accommodated. Flexible couplings, compression couplings, or other gasketed sleeve devices are used. Gasket materials should be suitable for long-term exposure to high temperatures. Similarly, seating materials on valves should also be suitable for high-temperature service. In specifying aeration system piping, the manufacturer should be fully informed as to the conditions of service anticipated both while the system is in operation and when it is out of service. Careful attention must be paid when using PVC piping in a system with high-discharge pressures to guard against melting and instability. In a system with high discharge pressure, only materials that have been stress rated at or higher than the temperature involved should be used. Some applications may warrant consideration of cooling the air. Consideration also should be given to worst case scenarios, such as a tank being dewatered (nearly empty) and the drop pipe exposed, thus exposing the diffusers to high temperatures. Long-term structural and mechanical integrity and long-term maintainability will be greatly affected by the consideration given during design to the static and dynamic forces and temperature that the air distribution system is expected to withstand.

Installation and Operation Considerations. During construction and before plant startup, diffusers should be kept clean. Piping and holders should be cleaned before diffusers are set. Diffusers should not be connected to the air laterals nor placed in their holders until the plant is ready for service. After diffusers are installed, tests should be performed for leaks and for installation integrity by covering diffusers with a few inches of water and by applying air so that each diffuser passes air at a low rate. Following this, the diffusers should be submerged by 3 to 4 feet of water and the aeration pattern observed for even distribution. Some recommend quantitatively measuring uniformity within the individual diffusers and within the system as a whole, by performing inverted graduate and/or bucket catches when the system is submerged several inches.

When using porous diffusers, either in a new installation or when retrofitting, the air piping should be thoroughly cleaned before startup. When the air supply is interrupted, backflow of water will occur and the piping will fill with process water. Forcing process water in the lateral air piping back through a porous diffuser is a concern. Porous diffuser manufacturers have eductor purge systems that limit the amount of process water that is blown back through the system, but they do not provide total protection. With porous diffusers, an efficient air-filtering system should be used and air supply should be continuous. If the air supply is interrupted, the system pressure following restarting will need to be compared against records to determine if diffuser cleaning is necessary.

Because of PVC'S high-thermal expansion and contraction, the piping system is extremely sensitive to temperature variation and construction techniques both during installation and when the basin is taken out of service and drained. The primary effort should be to construct the system with adequate flexibility to allow differential movement of components. Thermal expansion and contraction of PVC components can be minimized by the following practices:[38]

- keep water over diffuser elements when not in use;
- blow air through diffuser elements when not in use; and
- avoid draining the basins for inspection during periods of prolonged heat or cold.

SPECIAL CONSIDERATIONS FOR POROUS DIFFUSER SYSTEMS.
Porous diffuser systems require special considerations for process and inlet/outlet design, drainage and cleaning, power supply, and air filtration.

Process Design. The operating characteristics of fine pore diffusers are different than those of other oxygen transfer devices in terms of flow pattern created, effect of wastewater constituents on diffuser layout, and the air-flow rate operating range.[6,7]

These factors generally dictate that fine pore diffused aeration systems be designed with tapered aeration capabilities in tanks with high-aspect ratios. At a minimum, the diffuser density (i.e., the effective basin floor area per diffusion unit) should vary with the highest density near the tank inlet and the lowest at the tank outlet. The design should be capable of meeting expected variations in air-flow requirements, considering both variations in process oxygen requirements and alpha factors along the length of the aeration basin. It may also be desirable to section the diffusion system into grids, with independent air supply control to each grid. For example, a total of three grids might typically be provided in an aeration basin with a length-to-width ratio of 3 to 1 or greater.

Failure to provide proper aerator tapering in tanks with high-aspect ratios or in staged tanks can result in inadequate oxygen transfer capacity and low DO concentrations at the inlet end. Such conditions have been found to accelerate biofouling of fine pore diffusers and contribute to other process-related or operational problems.[41]

Inlet and Outlet Design. Because of the poor horizontal-mixing characteristics of diffusers placed in a full-floor coverage configuration, the potential for short-circuiting exists, as illustrated in Figure 4.18.[28] Therefore, careful consideration must be given to the design and location of inlet and outlet structures. Wastewater and return activated sludge should always be added across the entire width of an aeration basin. Good influent dispersement can be accomplished using weirs, multiple-inlet ports or a diffuser. Examples are shown in Figure 4.19.[28] Although not as critical as inlet conditions, outlet structures should also be designed to collect wastewater across the entire basin width to minimize short circuiting.

The point of entry of flow into an empty aeration basin should also be considered. For example, entry over a weir may be acceptable when the basin is full but may cause damage to the diffusers or header system

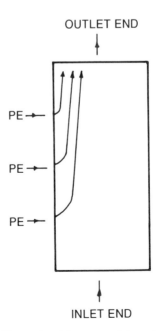

FIGURE 4.18—Short circuiting potential on fine bubble full floor coverage systems.

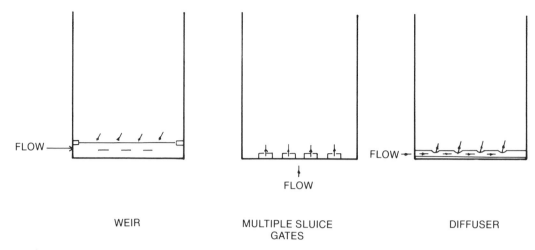

FLOW ⟶

WEIR

FLOW

MULTIPLE SLUICE
GATES

FLOW ⟶

DIFFUSER

FIGURE 4.19—Examples of acceptable aeration basin influent structures.

below when the basin is empty. In this case, an alternate means of filling the basin to a depth of at least several feet above the diffusers must be provided (perhaps through the basin drainage system).[7]

Basin Drainage and Diffuser Cleaning. For fixed porous diffuser systems, provisions must be made to drain the basins quickly and easily and to access the diffusers for maintenance. In addition, adequate facilities should be provided for on-site or *in-situ* diffuser cleaning. Facility requirements differ for the particular cleaning method chosen, but any system should have a good high- or low-pressure washdown system. Also, an adequate supply of spare parts should be stocked at the site. Enough diffusers to fully retrofit one basin, along with gaskets, hardware, and spare air supply piping should be stocked as a minimum.

Power Supply. The reliability of the power supply is a final consideration. Interruptions in service allow mixed liquor to enter the air header system through the diffusers and through any leaks in the system. Some suspended solids will be filtered out by the diffusers, but some will also enter the system piping. When air supply is resumed, a properly designed purge system should be used to clear the system so that the suspended solids will not be trapped and retained on the inner surfaces of the diffusers. Suspended solids filtered out by the diffusers during a power outage may be retained within the media on resumption of air delivery. These retained solids will result in higher headlosses across the diffusers and may lead to a change in OTE. Consequently, extra care should be taken during design to provide a reliable power supply with appropriate backup to minimize the occurrence and duration of power outages.[7]

Air Filtration

A properly filtered air supply is an important consideration for minimizing fouling of porous diffusers. Cleaning efficiency is the primary filter design characteristic. For fine pore diffusers, the common standard recommended for air quality has historically been 95% removal of particles 0.3 microns and larger. This "standard" may be in transition because of improvements in air main operation. Alternatively, specifications for filtered air quality require particulate concentrations less than 0.1 mg/ 29 cu m (0.1 mg/1000 cu ft).[1]

"In a retrofit, the requirements for filtration can be readily determined by conducting system air through a nonsubmerged test diffuser at several times the design flow and periodically measuring Dynamic Wet Pressure and Bubble Release Vacuum of upper and underside of test diffuser.

Types of Air Filters. Three basic types of air-cleaning systems are available: viscous impingement, dry barrier, and electrostatic precipitation. These filters are manufactured in a variety of forms and sizes with either manual or automatic operation.

Viscous impingement filters remove dust by impingement and retention of the particles on a labyrinth of oil-coated surfaces through which the air is passed. These units will handle a considerable quantity of dust without difficulty, and will effectively remove heavy particles. A large proportion of low-specific–gravity particles, however, will pass through in the air stream. Consequently, viscous filters are suitable for primary filtering. A coarse viscous filter ahead of an electrostatic filter can be a very good investment in a dusty area.

Dry barrier filters use a very fine filter material such as paper, cloth, or felt to entrap particles. The most common of these are in-line, replaceable filters. These systems are generally provided with a coarse fiberglass prefilter, followed by fine final filters. The prefilter normally consists of a roll of fiberglass cloth mounted on a frame. The final filters are housed in racks behind the prefilter. Replaceable filter systems take up little room, are easy to maintain, and are the simplest method to filter air. The units, however, are relatively expensive to replace. Large plants may wish to investigate bag house collectors. These units are constructed as steel enclosures that house sets of cloth-stocking tubes that are precoated with filter aid before being placed into service and after each cleaning. Their efficiency increases during a filter run because retained particles increase the effectiveness of the straining medium. A good dry filter, properly installed and maintained, will give protection up to the recommended standard if the atmosphere is not too smoky. The size, expense, and precoat requirements of bag filters have diminished their selection for newer plants.

Electrostatic precipitators impart an electric charge on particles so that they can be removed by attraction to elements of opposite polarity. Thse units require one third to one half of the area of bag houses and have relatively simple maintenance needs. Electrostatic filters will remove a large part of the very fine dust, including smoke, and will give protection up to the recommended standard when operated at velocities under 122 m/min (400 fpm). They are especially desirable for a smoky atmosphere.

Selection of Filter Types. Of the air-cleaning systems available, the replaceable filters are the simplest to construct and operate. Capital costs for these units are about one-eighth of electrostatic precipitator costs.[6] Bag house dust collectors are bulky and expensive, although apparently not maintenance intensive.

Replaceable air filters are the recommended approach except in cases where poor air quality requires replacement of the fine filtration elements more frequently than once per year. In this case, electrostatic units may be cost effective and should be considered.

Design Precautions. In addition to the design recommendations of the filter manufacturers, there are factors that require special attention for an activated sludge plant installation. Primarily, a wastewater treatment plant is expected to operate continuously 24 hours a day throughout the year; therefore, ample space and facilities for maintenance should be provided accordingly.

Weather protection of the air intake is important for continuous operation, especially in colder climates. Good louvers and an ample chamber between the louvers and filters are essential. In cold climates, preheating of the air is sometimes necessary to prevent snow or vapor from freezing onto and blanking the filters. When the intake is near open tanks of a treatment plant, the freezing vapor problem is especially bad. A simple method of preheating allows ducts and dampers to take a part of the air from inside the blower building when necessary. Locating the filter inlet should be done carefully to prevent sucking excessively moist air onto the filters; this can result in soaking the filter medium and reducing performance. The housing for air filters should be constructed from noncorrosive materials.

BLOWER SYSTEMS

A blower is a single or multiple-stage mechanical device used to produce relatively large volumes of air or gas at or near atmospheric pressure. A compressor can be classified as a single or multiple-stage device designed to produce lower volumes of air at higher pressures.[58,59] Although both machines may be used to perform some of the same functions, application and pressure range usually determines whether the device is called a blower or a compressor. Because their function may be the same, the terms blower and compressor at times are used interchangeably.

Among many wastewater treatment uses, blowers and compressors are used to provide oxygen and mixing for biological treatment, solids digestion, and combustion. They are also used, especially the positive displacement type, to provide an economical flexible source of power to operate pneumatic equipment, pumps, and tools, and for instrument air supply.

Blowers for aeration in activated sludge plants are the major source of energy consumption. Any attempt to minimize energy consumption here would be advantageous and could result in substantial cost savings. Proper application of a blower for aeration can result in high efficiencies. Annual electric power costs for blower operations are shown in Table 4.5.[60]

In the past, to vary the discharge from a blower, variable inlet vanes or outlet dampers were used. This mode of control provided the variable flow control needed at the expense of energy efficiency. Various other means are available to control discharge and are discussed later in this section.

Heat recovery is also available in low-pressure blower applications. When air is compressed, air temperature increases substantially. This temperature rise is typically 54°C at 55kN/m² (90°F at 8 psig) and 89°C at 110 kN/m² (160°F at 16 psig) for blowers that are 70% efficient. As blower efficiency drops, more waste heat is available.

Air to water heat exchangers that extract heat from the discharge air temperature can be used. Water temperatures exiting the heat exchan-

Table 4.5. Annual electric power costs for blower operation.

	Centrifugal	Positive displacement
Motor rating, kW (hp)	150 (200)	110 (150)
Motor efficiency, E	92%	90%
Brake horsepower (bhp)	155	140
Total daily operating hours	24	24
Unit power costs, R ($/kWh)	0.3578	0.3578
Annual electrical power costs[a] ($)	39 400	36 400

[a]Annual electric power cost $= \dfrac{\text{bhp} \times 24 \times 365 \times 0.746 \times R}{E}$

Note: $/kWh × 0.278 = $/MJ

ger can be as high as 60°C (140°F). This hot water is normally used to heat buildings or anaerobic digesters. A 65°C (150°F) blower discharge air temperature can provide approximately 10 W (1400 Btu/min) for each 480 L (1000 scfm) of air pumped. This corresponds to 2.8 L/h (0.75 gph) of fuel oil, or 5.6 L/s (1.4 cfm) of natural gas. If deep aeration tanks are used, additional heat is available because of the higher discharge pressures. An analysis of the system should be conducted to determine cost pay backs as well as effects of cooling the discharge air before the project is implemented.

Many times, costs can be reduced by properly maintaining the aeration diffusers. As diffusers plug, additional head loss is added to the system. The additional head requires a blower to do more work to accomplish the results. Discharge pressure readings should be logged periodically to determine when system problem analysis and correction should begin. Likewise, air leaks in the header system may be costing a substantial sum in energy costs. Corrections made in these two areas could greatly reduce energy costs.

THEORY OF GAS COMPRESSION. The compression of gas by blower or compressor follows natural physical laws. Many textbooks and handbooks[58,59] are available to provide detailed information on the gas compression processes. The WPCF manual on Prime Movers,[60] Manual of Practice No. OM-5, briefly discusses the most commonly used gas laws and formulas. The reader should refer to this literature for detailed discussions on gas compression.

In most wastewater treatment plants with diffused air systems, the operating pressure of the blowers is relatively low. The adiabatic compression process describes very closely the compression cycle of most positive displacement compressors and the dynamic blowers of low-compression ratio. Recently the compressor industry has used the polytropic process, which is a modification of the adiabatic process. Although the polytropic process is much closer to the actual compression cycle, it is not a true reversible process. For dynamic compressors with high-compression ratio, the polytropic process is a better choice than the adiabatic process.[58]

SYSTEM DESIGN CONSIDERATIONS

System design considerations include process requirements, environmental conditions, and systems analysis.

Process Requirements. Before designing an air supply system for the aeration system, the design engineer must first define the characteristics of the waste flow. The variation of the waste strength and the frequency of occurrence are important in establishing the air requirements. Chapter 2 discusses in detail the methods used to determine the required oxygen under standard conditions.

Environmental Conditions. The air-flow rates under standard conditions must be converted to various operating environmental conditions. The maximum waste load during a hot and humid day will determine the maximum air requirement. Similarly, the minimum waste load during a cold and winter night will set the minimum air requirement. The reader can obtain this seasonal weather information from the U.S. Weather Bureau. The conversion should include the correction for relative humidity.[59]

System Design Analysis. After conversion for environmental conditions, the design engineer should establish the air supply system configuration. Previous discussions on diffuser type, aeration reactor configuration, and piping layout provide valuable information. The designer will estimate a temperature rise across the compressor before calculating the pressure rise, and will adjust it afterward. These adjusted inlet and discharge pressures establish the required pressure rises for the blower or blowers. The designer then plots these calculated pressure rises and their corresponding actual air-flow rates to form a system demand curve. The system demand curve is essential in the analysis of blower operation.

TYPES AND CHARACTERISTICS OF BLOWERS. The positive displacement and the dynamic compressors are the two common types found in a wastewater treatment plant. Each type includes many different configurations that have their own variations in characteristics and in constructions. Positive displacement compressors are machines that successively compress a fixed volume of gas in an enclosed space to a higher pressure. The two most commonly used types are the rotary lobe compressors and the rotary helical screw compressors. The dynamic compressor moves the gas by its rotating impellers, which imparts velocity and pressure to the continuously flowing gas. The two commonly used types are the axial compressors and the centrifugal compressors. The following discussions compare the performance characteristics of these compressors. The reader should refer to the Prime Mover manual for detailed descriptions of these compressors and blowers.[60]

PERFORMANCE CHARACTERISTICS. The performance characteristics of the positive displacement and dynamic compressors are distinctly different. A basic understanding is essential in the proper application of these compressors.

Positive Displacement Compressors. The positive displacement compressor is a machine of constant capacity with variable pressure. The capacity of its driver limits the maximum pressure supplied by the positive displacement compressor. Figure 4.20[59] demonstrates the characteristics of a positive displacement rotary compressor at a constant rotational speed. Line AB illustrates the theoretical performance curve of a positive displacement compressor at a constant speed. At higher pressure, a small gas flow will slip between the casing and the rotor. This internal slippage causes the slight reduction of flow volume as shown by line AC. For the same compressor at different rotative speeds, the characteristic curves will follow the lines DE and FG.

Dynamic Compressors. The dynamic compressor is theoretically a constant-pressure, variable capacity machine. Without any internal losses, the theoretical performance curve shall be a horizontal line AB as shown in Figure 4.21.[59] In reality, the curve dips as AC or AD, depending on the impeller design and compression ratio. The gas density affects the performance of a dynamic compressor. Any change in the inlet gas temperature and the barometric pressure will cause a

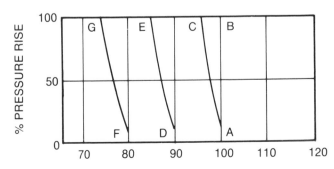

FIGURE 4.20—Characteristic curves of rotary compressors at different rotational speeds.

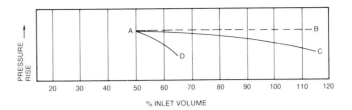

FIGURE 4.21—Characteristic curves of dynamic compressors. An increase in gas density causes an increase in pressure.

change in the density of the compressed gas. For a particular impeller and rotational speed, a change in gas density will change the pressure. Higher temperature and lower barometric pressure reduce the gas density. The greater the gas density is, the higher the pressure will rise. Denser gas requires greater brake horsepower for the compression process. Figure 4.22[59] shows the characteristic curves of a dynamic compressor of centrifugal type.

Differences Between Axial and Centrifugal Compressors. The performance curves are different between the axial and the centrifugal compressors. In a centrifugal compressor, the gas flows continuously along the flow path and forms a circular vortex by the rotating impeller. In an axial compressor, there is no vortex motion, and the gas velocity remains relatively constant. In Figure 4.23,[58] the axial compressor has a much steeper capacity versus pressure curve than the centrifugal compressor. The power versus capacity curve of the axial compressor slopes in the opposite direction of centrifugal compressor.

Differences Between Impellers of Centrifugal Compressors. Among the centrifugal compressors, the design of the impeller vane curvature affects their performances. The vanes are available in forward curved, radial, and backward-curved types. The exit angle of the gas from the tangent of the vane tip determines the type of the impeller. An exit angle of less than 90 degrees forms the forward curved impeller. A 90-degree, exit angle forms the radial impeller. An exit angle of more than 90 degrees forms the backward curved impeller.

Figure 4.24[58] shows the characteristic curves of the commonly used backward curved and radial impellers. The straight line AB represents

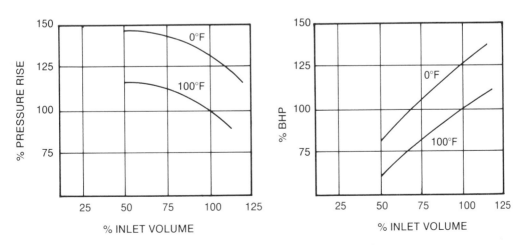

FIGURE 4.22—Characteristic curves of a centrifugal compressor. As gas density increases, greater power is needed for the compression process.

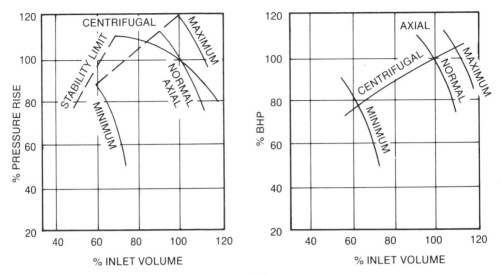

FIGURE 4.23—Comparision of axial and centrifugal characteristic curves.

the theoretical pressure versus capacity curve. Internal friction losses cause the line to curve as AC. Additional circulatory and shock losses move the curve lower as line DE. The point D is the limit of stable operation or the surge limit. As the flow falls below the surge limit, the compressor cannot produce enough discharge pressure to overcome the system pressure, resulting in a momentary flow reversal inside the compressor. The flow reversal causes the system pressure to drop, which restores the forward flow. This pulsating operation causes rise of discharge temperature, noise, and vibration inside the compressor.

ANALYSIS OF BLOWER OPERATION. In wastewater treatment plants, the blowers must supply a wide range of air flows with a relatively narrow pressure range under varied environmental conditions. A compressor usually can only meet one particular set of operating condition efficiently. Seldom can this particular performance satisfy the range of flows and pressures required for a wastewater treatment plant

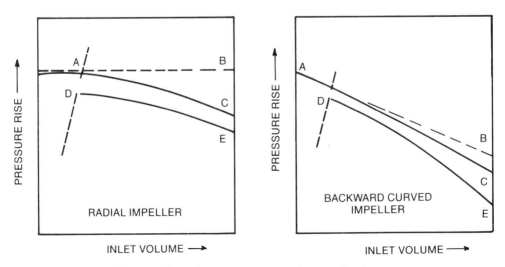

FIGURE 4.24—Effect of vane angle on characteristic curve of a centrifugal compressor.

operation. The operator must have a method to turndown the compressor capacity. The following discussion presents some of the commonly used methods to regulate or to turndown the compressors. A typical rating scheme for compressor or blowers is shown in Table 4.6.[60]

Flow Blowoff or Bypassing. For a positive displacement compressor, a blowoff valve on discharge or a bypass valve to intake are two simple ways of capacity turndown. These methods do not save any energy and should be used only on very small installations.[58]

Inlet Throttling. For the dynamic compressors, the most economical and efficient method of capacity turndown is inlet throttling. Inlet throttling should not be used on positive displacement compressors. A closed inlet valve may result in mechanical damage to the compressor. The throttling of the intake gas of a dynamic compressor will reduce the density of the inlet gas. This lowering of gas density causes reduction of the gas volume delivered at compressor discharge. The simplest method is the use of a butterfly valve on the compressor inlet. The more efficient method is the use of adjustable inlet guide vanes instead of the throttling valve. These devices prerotate the incoming gas, resulting in a reduction of both discharge capacity and pressure.

Figure 4.25[58] shows the effects of the inlet guide vanes on a backward-curved centrifugal compressor. Using the inlet guide vanes, the operator can lower the surge limit to nearly 30% of the rated capacity. Using an inlet-throttling valve, the operator can lower the surge limit to about 45% of the rated capacity.

Adjustable Discharge Diffuser. For a centrifugal compressor equipped with radial-type impeller, the adjustable discharge diffusers can provide capacity regulation without reduction in discharge pressure. These discharge diffuser vanes adjust the flow passage area ahead of the discharge nozzle without any impediment to the gas flow. Figure 4.26[61] illustrates the effects of adjustable discharge diffusers on the performance of a radial-blade centrifugal compressor. This device will have a surge limit at about 45% of the rated capacity. When working in conjunction with an adjustable inlet guide vanes, the combination will provide a very efficient capacity regulation system. The surge limit will be about 30% of the rated capacity.

Variable Speed Driver. The use of a variable-speed driver is an efficient method of capacity regulation for positive displacement compressors. For centrifugal compressors, the variable-speed driver is seldom used. A small change in rotative speed will produce a relatively large change in the discharge pressure of a centrifugal compressor. Because most aeration systems operate within a small range of pressures, the use of an expansive variable-speed driver is not economical. The Figure 4.27[59] presents the effects of variable-speed drivers to a positive displacement compressor and to a centrifugal compressor.

Parallel Operation of Multiple Units. For small wastewater treatment plants, a single blower with a standby and device for capacity turndown may be adequate. In a larger plant, the range of required operating capacity can be very large. In many cases, a single blower will not have the capacity to cover the wide range required. The design engineer must provide at least two or more blowers; one of which is a standby.

Figure 4.28 illustrates the analysis required to define the efficient operation of these multiple blowers. It depicts a typical parallel operation using two centrifugal blowers with adjustable guide vanes for capacity regulation. The system demand curve for the plant must first develop in accordance to procedures presented in the previous sections.

The maximum demand occurs on a hot and humid day of 100°F, (90% humidity). The combined performance curves of two identical

Table 4.6. Compressor ratings.

Class	Type	hp	Pressure (psig)	Volumetric (cfm)
Reciprocating positive displacement	[a]Piston—single stage	25–200	80–125	100–850
	Piston—multi stage	10–10 000	10–50 000	30–15 000
Rotary positive displacement	Lobe	10–3000	5–250	5–30 000
	Sliding vane	10–500	5–275	40–20 000
	Helical screw	10–500	10–250	20–15 000
Dynamic compressor	[a]Centrifugal	50–20 000	1.5–2000	20–25 000
	Axial	1000–10 000	40–500	800–13 000 000

[a]Certain light duty units may be available in smaller sizes and ratings.

NOTE: psig × 6.895 = kPa; cfm × 0.4719 = L/s; hp × 0.7457 = kW.

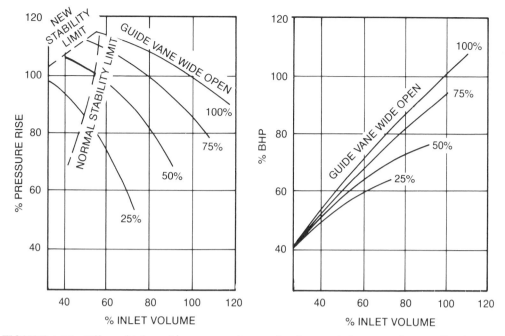

FIGURE 4.25—Effect of inlet guide vane rotation on the characteristic curve of a centrifugal compressor with backward curved impeller.

blowers at the above environmental conditions intersect the system demand curve at the required maximum demand point, A. As the air temperature drops to 0°F, the combined performance curve moves upward and intersects the system demand curve at point B. At this lower temperature, the two blowers move more volume of air at higher pressures, and they require more power for the operation. Because the cooler air contains more oxygen, there is no process need to supply this extra volume of air. Adjustments of the inlet guide vanes of both blowers to smaller openings will bring the point B down to point A. Assuming point A still supplies more than required oxygen, the operator can bring the performance curve down further to point C.

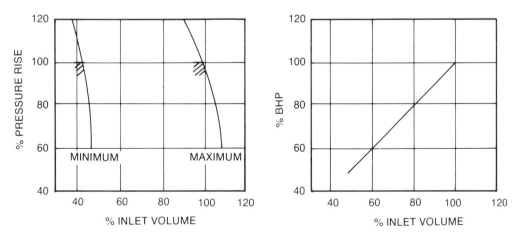

FIGURE 4.26—Effect of discharge diffuser rotation on the characteristic curve of a centrifugal compressor with radial impeller.

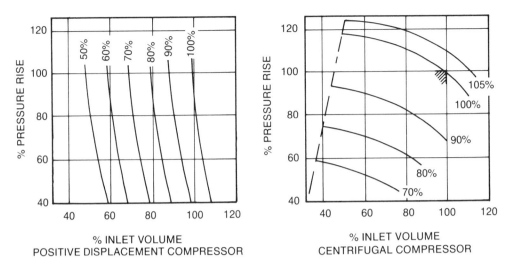

FIGURE 4.27—Effect of variable speed driver on the characteristic curve of a centrifugal compressor.

On a cold winter day of 0°F, a single blower with fully opened inlet guide vanes may not intersect the system demand curve. Adjustment of inlet guide vanes can bring the blower to its operable range as shown by the point D.

CONTROL OF SURGING. Surging in a centrifugal compressor occurs at a low volumetric range of operation. Below a particular minimum flow, a surge limit, the compressor performance is unstable. The pulsating surge produces noises, vibrations, and higher air temperature in the compressor. The following discussion presents the commonly used methods of surge control.

Blow-off Portion of Discharge Air. Installation of a blow-off valve on compressor discharge to vent the air to the atmosphere is a simple way to control surge.[58] Venting air will create noise and will raise the surrounding temperature. In small plants, this may be the only method used. In larger plants, however, it is commonly used as a back-up device.

Bypassing Portion of Discharge Air. Installation of a bypass valve to direct a portion of the discharge air to the compressor inlet is a simple method to control surge.[58] Bypassing into the inlet side contains the

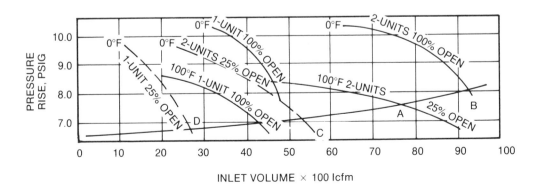

FIGURE 4.28—Parallel operation of two centrifugal compressors.

noises within the compressor and its piping. Continuously bypassing, however, will cause temperature build-up in the compressor, which requires cooling of the bypassed air.

Flow Regulation. Using either inlet valve or inlet guide vanes to keep the air flow above the surge limit is an efficient surge control method.[59] The operator should monitor either the mass flow rate of inlet air or the power input to the compressor. These monitored values will define the surge limit accurately. The discharge diffusers on a centrifugal compressor of radial impeller can also perform similar surge control function.

BLOWER START-UP. The start-up of a blower or a compressor requires many precautions. Start-up for a blower with a closed discharge valve can cause mechanical damage to the blower.[59] An unloading arrangement on the discharge side of a blower should be used during start-up. The unloading arrangement could be a blow-off or a bypass valve as described in surge control. On positive displacement compressors, a relief valve of discharge side will provide the extra margin of safety.

DRIVERS FOR THE BLOWERS. The selection of the type of the driver for a blower requires careful consideration of many factors: (a) the capacity of the driver must meet the power requirement of the blower; (b) the availability and the economy of the electric power, liquid or gaseous fuel that determines the selection of the driver; and (c) ease of operation and maintenance as part of the life-cycle cost consideration. The reader should refer to the Prime Mover manual for additional discussions on engines, motors, and turbines.[60]

TESTING FOR PERFORMANCE. Performance testing of a blower under a controlled environment is the best method to establish its meeting of required conditions. The ASME Power Test Code (PTC) Committee has established detailed test procedures for the blowers and compressors. ASME PTC-9 applies to the positive displacement compressors, and ASME PTC-10 applies to the dynamic compressors.

BLOWER INSTALLATION

Layout. Installation requires full consideration of layout requirements, foundation needs, and piping. Efficient operation of compressors and their drivers depends on the existence of a well-planned installation. Compressors are precision machines. A clean, dry, and nonhazardous environment is essential for the compressors. Arranging the foundations, piping, controls, and all of the auxiliary appurtenances should provide efficient operational conditions. Ample spaces should be provided around equipment to ease the installation and the dismantling of the compressor and its auxiliaries. Easy access to critical parts is essential for inspection and for maintenance. Valves and controls should be easily accessible. Hoisting equipment is a necessary facility required not only for installation but also for periodic maintenance.[59]

Foundation. The foundation of a compressor must be able to keep the compressor in alignment with its driver. The design must minimize the transmission of vibration to surrounding structures. The loading should be properly distributed to reduce large uneven loading. Loading must be within the safe dynamic bearing capacity of the supporting soil.[59]

Reciprocating compressors require careful study of foundation requirements. The resultant force of the vertical load and the unbalanced horizontal reciprocating forces should be within the base area of the foundation to avoid tilting and toppling. The mass and the bearing area must be large enough to avoid sliding because of unbalanced forces.

The high-rotative speed of the rotary and the dynamic compressors demand different requirements on the foundation design. A rigid and nonwarping support is critical to the dynamic compressors. Unequal settlement of the baseplate will result in misalignment of the compressor and its driver. For large heavy machines, vibrations must be isolated from the surrounding structures. This requires a large mass foundation, rigid structure support, and the use of vibration isolators under the machines.

Piping. Heat generated by compression raises the discharge temperature. This high heat subjects the discharge piping to large variations of temperature. The thermal expansion and contraction experienced by the discharge piping are the main causes of piping strain. The piping should be arranged to avoid strains that cause misalignment of the compressor and the piping. Control of these thermal strains requires the use of expansion joints. Flexible-type expansion joints on blower inlet and discharge connections should be installed. This also reduces the transmission of vibration between the compressor and its piping.[59]

NOISES AND VIBRATIONS. Noises and vibrations are common phenomena for all rotating machines. They are generated by the rotating element of the compressor and its components, by the geared speed increaser, and by the driver. All of these elements interact with each other, with the connecting piping, and with its mounting structure. The transmission of sound is through air, structure, or both. The transmission of vibration is through direct connections among the equipment, piping, and structure. The Walsh-Healey Act on Occupational Noise Exposure sets the legal limits of permissible noise exposure for various sound levels. The design engineer may wish to lower the noise level below these legal limits for human comfort.

Control of Noises. Noises generated inside the compressor are transmitted through its intake and discharge openings. Dynamic compressors rotating at high speed produce higher frequency sounds than at low speed. A positive displacement compressor generates lower frequency sounds. The characteristics of the noises will determine the selection of the type of silencer.

Internal-gearing noises radiated from the housing can be attenuated by treating with damping material. The electric motor generates noises by its cooling fans and by the windage of its rotating elements. These noises can be disturbing from a high-speed, large horsepower motor. The use of a weather-proof enclosure with internal acoustic baffles will provide effective noise attenuation.

Noises radiating from piping carrying high velocity air flows can be annoying. Heavy walled pipe or pipe lagging should be used to control these piping noises. Rigid piping supports are essential to combating the noise generation. Flexible expansion joints between the compressor and the piping will reduce the noises caused by the transmitted vibrations. Using sound-absorbing materials on the walls controls the reverberated sound within the blower room.

Vibration Control. A high-speed rotating machine has a critical speed. At this critical speed, the machine vibrates violently. Vibration may be lateral or torsional. The rotative speed of compressors should be at least 20 to 30% below or above the critical speed.[61] Design of the rotating shaft should include the critical speed in its stress analysis. A dynamic balance of the rotating element, a careful alignment of the machine, and good piping and foundation design are essential in achieving smooth operation.

REFERENCES

1. "Aeration in Wastewater Treatment." Manual of Practice No. 5, Water Pollut. Control Fed., Washington, D.C. (1971, 1952).
2. Roe, F.C., "The Installation and Servicing of Air Diffuser Mediums." *Water and Sewer Works,* **81,** 1934, 115–123.
3. Beck, A.F., "Diffuser Plate Studies." *Sewer Works Journal,* **8,** 1936, 22–37.
4. Committee on Sewage Disposal, "The Operation and Control of Activated Sludge Sewage Treatment Works." *Sewer Works Journal,* **14,** 1942, 3–69.
5. Boyle, W.C. and Redmon, D.T., "Biological Fouling of Fine Bubble Diffusers: State-of-Art." *ASCE Journal of Environmental Engineering,* **109,** 5, 991–1005 (1983).
6. Houck, D.H. and Boon, A.G., "Survey and Evaluation of Fine Bubble Dome Diffuser Aeration Equipment." EPA-600/2-81-222, Municipal Environmental Research Laboratory, U.S. EPA, Washington, D.C. (1981).
7. U.S. EPA and ASCE, "Summary Report: Fine Pore (Fine Bubble) Aeration Systems." EPA-625/8-85-010, Water Engineering Research Laboratory, U.S. EPA, Washington, D.C. (1985).
8. Bartholomew, G.L., "Types of Aeration Devices." In: *Aeration of Activated Sludge in Sewage Treatment.* D.L. Gibbon, (Ed.) 25–27, Pergamon Press, Inc., New York (1974).
9. Winkler, W.W., "Fine Bubble Diffuser Maintenance." Presented at the annual meeting of the New England Water Pollution Control Association, Boston, Mass., January 25, 1984.
10. Ewing, L. and Redmon, D.T., U.S. Patent No. 4,261,933 (April 14, 1981).
11. Toerber, E.D. and Mandt, M.G., "Greater Oxygen Transfer with Jet Aeration System." *Water and Sewage Works,* 1,71 (Jan 1979).
12. McCarthy, J.J., "Technology Assessment of Fine Bubble Aerators." EPA-600/2-82-003, Municipal Environmental Research Laboratory, U.S. EPA, Washington, D.C. (1982).
13. Personal communications from G. Powell, Gore & Storrie, Ltd., Toronto, Ontario, to B.R. Willey, CWC-HDR consultants, (1985).
14. Speece, R.E., Gallagher, D., Krick, C., and Thompson R., "Pilot Performance of Deep U-Tubes." Presented at the annual conference on Environmental Engineering, ASCE, New York, N.Y. (July 1979).
15. Robinson, M.S., "A Pilot-Plant Evaluation of a Deep Shaft Linked to Conventional Aeration Without Inter-Stage Settlement to Give a Fully Nitrified Effluent." *J. Water Pollut. Control Fed.* **85,** 58–70 (1984).
16. Bewtra, J.K., nd Nicholas, W.R., "Oxygenation from Diffused Air in Aeration Tanks," *J. Water Pollut. Control. Fed.* **36,** 10, 1195–1224 (1964).
17. Personal communications from Lloyd Ewing, Ewing Engineering Company, Milwaukee, Wis. to Ross McKinney, University of Kansas, Lawrence (1985).
18. Barnhart, E.L., "Transfer of Oxygen in Aqueous Solutions." *ASCE Journal of the Sanitary Engineering Division,* **95,** 3, 645–661 (1969).
19. Personal communication from D. Redmon, Ewing Engineering Company, Milwaukee, Wis. to B.R. Willey, CWC-HDR consultants, Seattle, Wash. (1985).
20. Danley, W. "Fouling Study of Fine Bubble Ceramic Diffuse Phase I." MS Report, University of Wisconsin, Madison, Wis. (1984).
21. Huibregtse, G.L., Rooney, T.C., and Rasmussen, D.C., "Factors

Affecting Fine Bubble Diffused Aeration." *J. Water Pollut. Control Fed.* **55,** 8, 1057–1064 (1983).

22. Sullivan, R.C. and Gilbert, R.G., "The Significance of Oxygen Transfer Variables in the Sizing of Dome Diffuser Aeration Equipment. Presented at 56th Annual WPCF Conference, Atlanta, Ga. (1983).

23. Yunt, F., Hancuff, T., Brenner, R., and Shell, G., "An Evaluation of Submerged Aeration Equipment—Clean Water Test Results." Presented at the 1980 WWEMA Industrial Pollution Conference, Houston, Tex. (1980).

24. Gerry Shell Environmental Engineers, Inc., "Final Report—Oxygen Transfer and Headloss Characteristics of the Carborundum Dome and Tube Fine Bubble Diffusers." (1980).

25. Paulson, W.L., "Oxygen Absorption Efficiency Study—Norton Co. Dome Diffusers." P.F. Morgan Laboratory, Iowa City, Iowa, (1976).

26. Personal communications from F. Yunt, Los Angeles County Sanitation Districts, to B.R. Willey, CWC-HDR consultants, Inc. (1985).

27. Tomlinson, E.J., and Chambers, B., "The Effect of Longitudinal Mixing on the Settleability of Activated Sludge." Water Research Centre, Technical Report, TR 122, Stevenage, U.K. (1979).

28. Wyss Flex-A-Tube Diffuser Systems. Product information bulletin, Wyss, Inc., Ada, OH (1983).

29. Rooney, T.C. and Huibregtse, G.L., "Increased Oxygen Transfer Efficiency with Coarse Bubble Diffusers." *J. Water Pollut. Control Fed.* **52,** 9, 2315–2326 (1980).

30. Schmit, F.L., Wren, J.D. and Redmon, D.T., "The Effect of Tank Dimensions and Diffuser Placement on Oxygen Transfer." *J. Water Pollut. Control Fed.,* **50,** 1750–1767 (1978).

31. Mueller, J.A. and Saurer, P.D., "Field Evaluation of Wyss Aeration System at Cedar Creek Plant, Nassau County, N.Y." (August, 1986).

32. Yunt, F.W., "Aeration Equipment Evaluation Phase II: Process Water Results." Draft final report for Contract 68-03-2906 between U.S. EPA and Los Angeles County Sanitation Districts.

33. Setter, L.R., "Air Diffusion Problems at Activated Sludge Plants." *Water and Sewage Works,* **95,** 450–456 (1948).

34. Anderson, N.E., "Tests and Studies on Air Diffusers for Activated Sludge." *Sewage and Industrial Wastes,* **22,** 461–476 (1950).

35. Lamb, M., "Designing and Maintaining Porous Tube Diffusers." *Wastes Engineering,* **25,** 405–413 (1954).

36. Wisley, W.H., "Summary of Experience in Diffused Air Activated Sludge Plant Operation." *Sewer Works Journal,* **15,** 909–935 (1945).

37. Addison, G. "Slime Growth on Fine Bubble Diffusers." Report by Gore & Storrie, Ltd., Toronto, Ontario, Canada.

38. CH₂M-Hill, Corvalis, OR "Dome Diffuser Evaluation." In-house memorandum. February 13, 1981.

39. Yunt, F.W., "Some Cleaning Techniques for Fine Bubble Dome and Disk Aeration Systems." Internal Rept., L.A. County San. Dist., L.A., CA (1984).

40. ASCE Committee on Oxygen Transfer, U.S. EPA Agreement 812167, Amer. Soc. Civil Eng., New York, N.Y. (1985).

41. Houck, D.H., "Survey and Evaluation of Fine Bubble Dome and Disc Aeration Systems in North America." Draft final report for Purchase Order No. C2667NASX between U.S. EPA and D.H. Houck Associates, Inc.

42. American Society of Civil Engineers. "ASCE Standard Measurement of Oxygen Transfer in Clean Water." ISBN 0-87262-430-7, New York, N.Y. (1984).

43. Mueller, J.A.,"Nonsteady State Field Testing of Surface and Dif-

fused Aeration Equipment." Manhattan College, Bronx, N.Y. (1983).

44. Redmond, D.T., Boyle, W.C., and Ewing, L., "Oxygen Transfer Efficiency Measurements in Mixed Liquor Using Off-Gas Techniques." *J. Water Pollut. Control. Fed.* **55**, 11, 1338–1347 (1983).

45. Stenstrom, M.K. and Gilbert, R.G., "Effects of Alpha, Beta, and Theta Factor Upon the Design, Specification and Operation of Aeration Systems." *Water Research,* **15**, 6, 643–654 (1981).

46. Popel, J.H. "Improvements of Air Diffusion Systems Applied in the Netherlands." In: *Proceedings of Seminar Workshop on Aeration System Design, Testing, Operation, and Control,* EPA-600/9-85-005, NTIS No. PB85-173896/AS, U.S. EPA, Cincinnati, Ohio, 156–176 (1985).

47. Paulson, W.L. and Johnson, J.K., "Oxygen Transfer Study of FMC Pearlcomb Diffusers." Report prepared for the FMC Corporation, Lansdale, Pa. (1982).

48. Personal communications from W.L. Paulson, University of Iowa, Iowa City, Iowa, to W.C. Boyle, University of Wisconsin, Madison, Wis., 1985.

49. Smith, D.W. "Aeration System Design Protocols: A North American Perspective." In: Proceedings of Seminar Workshop on Aeration System Design, Testing, Operation and Control, EPA-600/9-85-005, NTIS No. PB85-173896/AS, U.S. EPA, Cincinnati, Ohio (1985).

50. Lister, A.R. and Boon, A.G., "Aeration in Deep Tanks: An Evaluation of a Fine-Bubble Diffused-Air System." *Journal Institute of Water Pollution Control,* **72**, 5, 590–605 (1973).

51. Campbell, H.J., Jr., "Oxygen Transfer Testing Under Process Conditions." In: *Proceedings of Seminar Workshop on Aeration System Design, Testing, Operation, and Control.* EPA-600/9-85-005, NTIS No. PB85-173896/AS, U.S. EPA, Cincinnati, Ohio, 345–363 (1985).

52. von der Emde, W. "Aeration Developments in Europe." In: *Advances in Water Quality Improvement,* E.F. Gloyna, and W.W. Eckenfelder, Jr., (Eds.). University of Texas Press, Austin, Tex., 237–261 (1968).

53. Yunt, F.W., and Hancuff, T.O., "Relative Number of Diffusers for the Norton and Sanitaire Aeration Systems to Achieve Equivalent Oxygen Transfer Performance." Internal Rept., L.A. County, San Dist., L.A., Ca (Dec 14, 1979).

54. Personal communications from G. Huibregtse, Rexnord Corporation, Milwaukee, Wis, to B.R. Willey, CWC-HDR consultants, (1985).

55. Owen, W.F., (Ed.) "Energy in Wastewater Treatment." Prentice-Hall, Inc., Englewood Cliffs, N.J. (1982).

56. Spink, L.K., "Principles and Practice of Flow Meter Engineering." (9th Ed.), Foxboro Company, Foxboro, Mass. (1967).

57. Yunt, F.W., "Gas Flow and Power Measurement." In: *Proceedings of the Workshop Toward an Oxygen Transfer Standard,* EPA-600/9-78-021, Washington, D.C. (1979).

58. Loomis, A.W., (Ed.) "Compressed Air and Gas Data." Ingersoll-Rand Company, Woodcliff Lake, N.J. (1982).

59. "Compressed Air and Gas Handbook." Compressed Air and Gas Institute, New York, N.Y. (1966).

60. "Prime Movers—Engines, Motors, Turbines, Pumps, Blowers & Generators," Manual of Practice OM-5, Water Pollut. Control Fed., Washington, D.C. (1984).

61. Church, A.H., "Centrifugal Pumps and Blowers." John Wiley & Sons, Inc., New York, N.Y. (1944).

Chapter 5
Mechanical Aerators

INTRODUCTION

Major mechanical aerator applications include various activated sludge process modifications, including oxidation ditches and ring or oval-shaped treatment units, and aerated lagoons. Mechanical aerators are also often used in pre-aeration systems, equalization basins, post-aeration systems, polishing ponds and for the aeration of natural water bodies.

The following are the generally accepted major requirements for mechanical aerators:

1. Sufficient oxygen transfer must occur at a reasonable cost. The criteria that define this requirement are standard aeration efficiency, N_o, or the field aeration efficiency, N, defined as

   $$N_o = SOTR/P$$
   $$N = OTR/P$$
   where;
   N_o and N are expressed in kg/kWh, SOTR is standard oxygen transfer efficiency, kg/h, OTR is field oxygen transfer efficiency, kg/h, and P is net aerator power, kW.

2. Bulk liquid, mixed liquor solids, and oxygen mixing by convection and turbulence must be sufficient. Existing mixing criteria are not well quantified. The most common criterion is the specific energy requirement for mixing, generally defined as power required per unit volume of liquid. This criterion, however, is equipment- and facility-specific. Therefore, pertinent information must be obtained from equipment manufacturers and operating plants.

3. Aerosols and mist caused by aeration should not produce air pollution and not result in mist freezing. This requirement is important for aeration device selection when facilities are close to residential zones or in cold climate areas.

4. Aerators should require low maintenance and should be capable of providing operational and control flexibility.

Certain requirements are more important for particular applications. For example, sludge suspension and the associated mixing requirements in activated sludge processes are usually satisfied when aerators are selected on the basis of oxygen transfer. However, for aerated lagoons, ring or oval-shaped structures, and equalization basins, sludge suspension and the associated mixing usually determine aeration equipment size selection.

Motor-driven, mechanical aerators provide a combination of liquid aeration and mixing. Some mechanical aerators produce the gas-liquid interface by entraining air from the atmosphere and dispersing it into bubbles. Others disperse liquid in the form of droplets or they produce jets or thin films that contact the ambient air. Yet others generate both air bubbles and liquid droplets. A specific group of aerators makes use of diffusion and mixing by injecting air into the bulk liquid. Such aerators are called combined diffused-air mechanical aerators, and are sometimes referred to as "spargers."

A detailed mechanical aerator classification based on major design features and operating principles is presented in Figure 5.1. Mechanical aerators are usually divided into two major groups: aerators with a vertical axis and aerators with a horizontal axis. Both groups are further subdivided into surface and submerged aerators. Each subdivision includes several generic aeration devices. Major design features and operating principles, examples of aerators available on the market, and the most appropriate use of various generic aerator types are discussed in the following sections.

SURFACE MECHANICAL AERATORS WITH VERTICAL AXIS

Vertical-axis, mechanical-surface, aerator rotors can induce either updraft or downdraft flows below the aerator. Aerator impellers (also referred to as rotors) function as pump rotors and can be characterized by specific speed, n_s, which can be determined by Equation 1:

$$n_s = 3.65 \frac{n \, Q^{0.5}}{H} \tag{1}$$

Where:

n_s = specific impeller speed,
n = rotational impeller speed, min^{-1},
Q = flow rate through the impeller, L/s, and
H = hydraulic head produced by the impeller, m H_2O.

Usually n_s values are between 60 and 120, 120 and 350, and 400 and 800, respectively, for centrifugal, radial-axial, and axial impellers. Accordingly, centrifugal impellers belong to the low-speed devices, and the axial impellers belong to the high-speed devices. The low-speed impellers are usually driven by a motor through a gear box for speed reduction, while the high-speed impellers are usually directly driven by the motor output shaft. The radial-axial impellers have intermediate characteristics between centrifugal and radial devices; they are usually driven with gear boxes that have a lower reduction ratio than those typical for centrifugals.

UPDRAFT AERATORS. Principle schematics of updraft aerator flow characteristics are presented in Figures 5.2 and 5.3. These aerators can have either centrifugal (also called low-speed), radial-axial (also called medium-speed), or axial (also called high-speed) impellers.

CENTRIFUGAL (LOW-SPEED) AERATORS. Centrifugal impellers of surface vertical-axis aerators differ in design (Figures 5.2a and 5.2b). Some impellers have a flat disc with rectangular or slightly curved vanes attached at the peripheral section of the disk's lower sur-

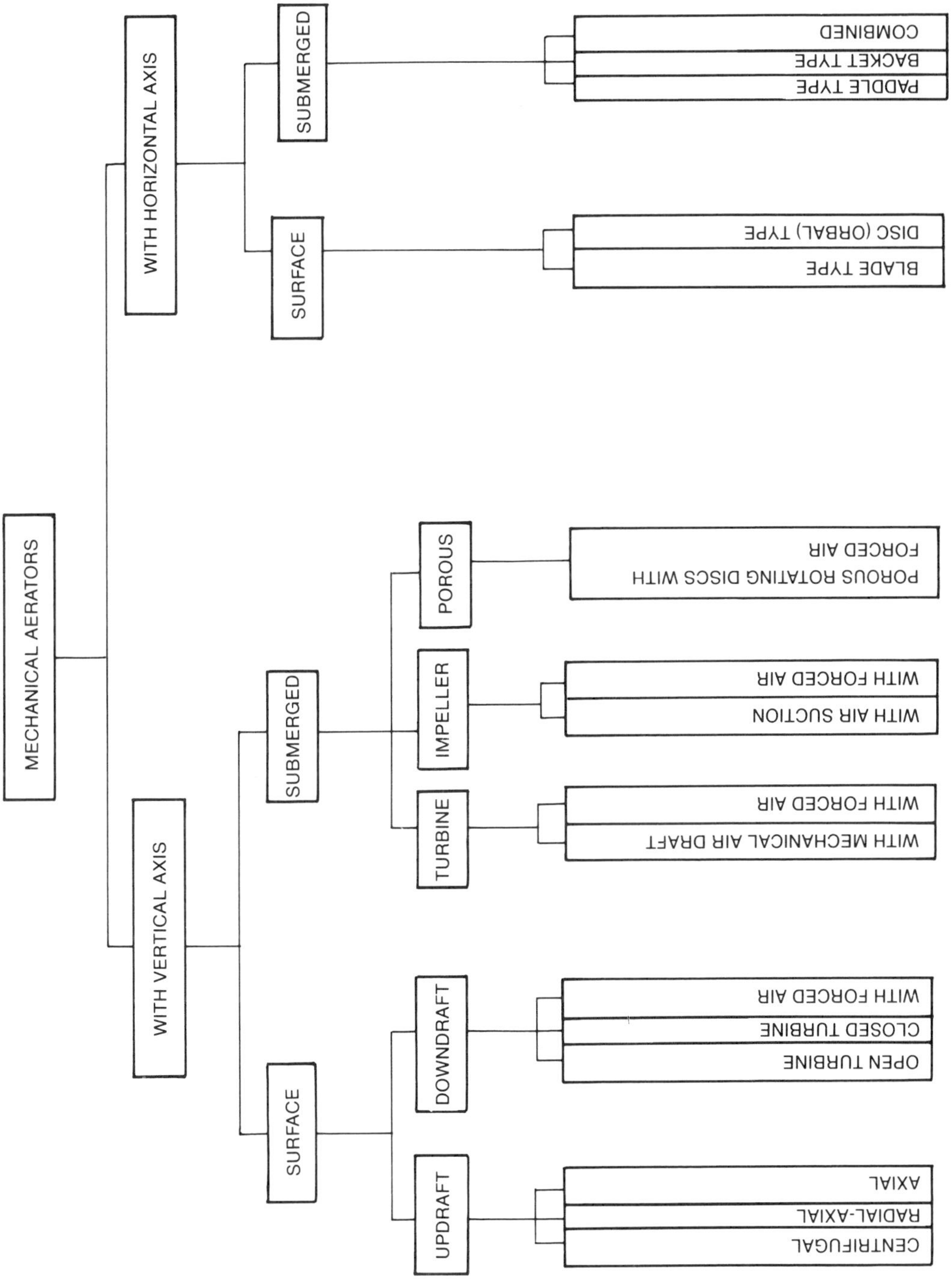

FIGURE 5.1—Classification of Mechanical Aerators.

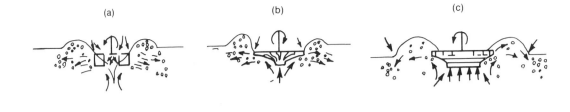

FIGURE 5.2—Flow characteristics for Updraft Vertical Axis Surface
        Mechanical Aerators; (a) and (b) centrifugal rotors, (c)
        radial-axial rotor, (d) axial rotor, horizontal discharge,
        (e) axial rotor, downward discharge.

face along, or at a slight angle to, the radii. As an option, holes for air
suction can be provided in the disc behind the vanes. The impeller is
usually completely open, or sometimes located inside a conical shroud,
and can be used with or without a draft tube.

Liquid is introduced to the aerator impeller in a predominantly vertical direction, then accelerated by the impeller vanes, and finally discharged in an essentially horizontal direction at the impeller rim. The
discharged flow can be considered as rapid (i.e., supercritical) flow and
slow-moving liquid in the vessel can be considered as tranquil (i.e.,
subcritical) flow. Transition from rapid flow to tranquil flow occurs
through a hydraulic jump. Therefore, a circular hydraulic jump is
formed around the aerator where the ambient air is entrained and dispersed into bubbles in the hydraulic jump. Simultaneously, bulk liquid
circulation and mixing is induced. The circulation patterns are determined by four major components: vertical circulation of upflow below
the impeller; surface horizontal flow from the impeller to the vessel
walls, downflow along the vessel walls, and the bottom horizontal flow
toward the aerator location. The bulk liquid generally rotates in the
direction of the impeller rotation while the flow regime is determined
by vessel geometry.

Mechanical aerators with centrifugal impellers have a gear-driven,
turbine-type impeller mounted horizontally just below the liquid surface
that rotates to draw large quantities of atmospheric air into the liquid.
Some examples of centrifugal impellers used as surface aerators appear
in Figure 5.3. The impeller is a circular flat plate with a series of
equally spaced radial blades extending downward. Behind the blades
are air suction holes. For additional hydraulic stability and clogging
prevention, impellers greater than 7.5 kW (10 hp) have a centrally
located, streamlined shape on the underside of the horizontal plate.
The shaft and impeller are steel and their entire weight is suspended
from the drive unit. Variable mounting is available in the form of a
fixed platform, a walkway, or a raft. Raft-mounted units allow for wide,
liquid-level variations and for periodic aerator towing for inspection.
For efficient mixing and aeration, pontoons are designed to eliminate
interference with generated hydraulic patterns.

Other centrifugal aerators include those having an inverted conical
body with blades originating at the cone center, an impeller in the

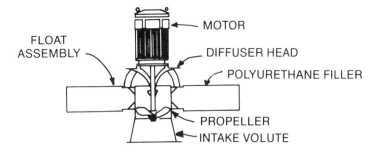

FIGURE 5.3—Mechanical surface aerator.

form of a centrifugal pump with curved vanes; and a disk-type impeller with radial blades both at the bottom and the top (Figure 5.4). The top blades prevent rotor flooding at greater submergence and ensure the optimal position of the circular hydraulic jump at the disk rim. This also reduces the wave generation in the aeration tank that causes splashing and pulsating loads on the gear-motor assembly. These pulsations often damage bearings and gears.

Aerators with centrifugal impellers can have a motor power ranging from several kilowatts to greater than 100 kW (140 hp). This type of aerator has been used successfully at numerous municipal and industrial wastewater treatment plants around the world for activated sludge processes, aerated lagoons, and ponds.

RADIAL-AXIAL AERATORS. Radial-axial vertical surface aerators are comprised of an impeller and a motor-gear assembly (Figure 5.2c). Typical impellers are either inverted cone bodies with curvilinear vanes or open blades attached to a circular frame or a drive shaft. These impellers can be used with or without a shroud and draft tube. The rotational speeds of radial-axial aerators are greater than that of centrifugal aerators, although the gear reducers are smaller and the aerators are lighter.

The flows discharged from these aerator impellers have radial horizontal and upward vertical components. These flows exit from the liquid mass in the aerated vessel, partially break into jets and droplets, and impinge into the aerated vessel liquid at a distance from the rotor. The ambient air is then entrained and dispersed into bubbles. Therefore, oxygen transfer is caused by both droplet and bubble formation. The water circulation patterns are similar to those generated by centrifugal impellers, although the intensity of the liquid mass rotation in the aerated vessel is less pronounced.

A typical aerator with radial-axial rotor having an inverted cone with radial self-cleaning blades is shown in Figure 5.5. These aerators typically have unit power from 7.5 to 110 kW (10 to 150 hp). The

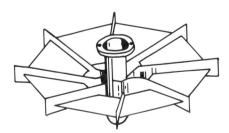

FIGURE 5.4—Mechanical aerator with disc-type impeller and radial blades.

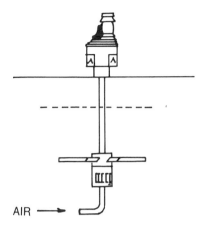

FIGURE 5.5—Radial axial submerged-turbine aerator.

aerator can be mounted on a fixed platform, walkway, or a raft. Stationary steel baffles built in a circular draft tube are located under the impeller for better flow distribution within the aerated vessel.

Other aerators with power ratings of 4 to 19 kW (5 to 25 hp) or 30 to 85 kW (40 to 125 hp) (Figure 5.6) have a disc impeller with individual blades attached at the disc periphery. The use of individually mounted blades provides greater flexibility when changing the unit capacity on the job site at a future date. The unit is driven by a standard gear motor assembly and is adaptable for the use of a variable-speed drive. The aeration unit can be installed on circular floats, on a rigid platform, or on a walkway. A drive-ring hood for preventing ice accumulation in all but extreme cold weather conditions is available.

A surge ring with flat baffles prevents wave generation in the tank and eliminates the pulsating loads on the gear-motor assembly thereby extending the life of the gear and bearings. The surge ring is also used as a draft tube, which provides improved oxygen distribution in the tank. Impeller and blade geometry ensure a high-hydraulic efficiency and the capability to pump large quantities of liquid at low head.

Aerators (Figure 5.7) with impellers having open arcuate blades originating at the drive shaft have the blade height which is reduced stepwise from the central section of the outer ring. Aerators on a rigid support structure can be equipped with a conical shroud and a draft tube. Floating aerators are usually supported by three floats and a lightweight tubular structure so that the entire system has a low center of gravity and therefore high stability. Floating aerators are not used with draft tubes. Mist shrouds for prevention of ice formation are provided. The unit power of these aerators is between 4 to 75 kW (5 to 100 hp). High flow rates delivered by the aerator, and the draft tubes when needed, may provide efficient sludge mixing and suspension.

AXIAL TYPE (HIGH–SPEED) AERATORS. Axial-flow, vertical-axis aerators (Figure 5.2, e and d) have a propeller-type impeller driven by a motor without a gear box, a shroud in which the impeller is located, and a flow-directing casing. The design of this casing determines the direction of the liquid jets discharged from the aerator: upward and away from the aerator, horizontal from the aerator, and, downward and away from the aerator. The discharge jets partially break into droplets, then entrain and disperse the air into bubbles on impingement into the bulk liquid in the vessel. The large surface area of droplets and bubbles ensures effective OTR. Large-flow rates developed by the aerator impeller provide substantial mixing amount and

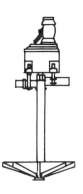

FIGURE 5.6—Submerged low-speed aerator.

biomass suspension. Flow patterns induced by these aerators in the vertical plane are similar to those induced by centrifugal impellers; however, general bulk liquid rotation within the vessel is virtually absent.

One type of high-speed aerator (Figure 5.8) has a claimed power range from 2.2 to 19 kW (3 to 25 hp). These aerators are mounted on a ring-shaped float.

Another type of aerator which is similar to the kind previously described is available in a wider power range from 2.2 to 120 kW (3 to 150 hp) with additional options such as stainless steel, single floats for smaller units or an assembly of three floats for larger units, draft tubes and anti-erosion accessories, and a cold-weather accessory package.

DOWNDRAFT AERATORS. Downdraft surface aerators can be classified as open-turbine, closed-turbine, and forced air-propeller aerators.

Open-Turbine Aerators. Open-turbine, downdraft surface aerators have a bladed impeller on a vertical shaft and a motor. Ambient air is entrained by the downward flow induced by the impeller, and then is dispersed into bubbles. At the same time, the downward flow causes bulk liquid mixing. The water circulation patterns are: vertical circulation towards the aerator at the surface; downward below the aerator; towards the vessel walls at the bottom; and, upward along the vessel walls.

One example of an open-impeller, downdraft aerator (Figure 5.9) has a motor-gear drive and a vertical-axis, truncated cone impeller with spiral vanes that induce radial-axial flows.

Closed Turbine Aerators. Closed-turbine, downdraft-surface aerators have an aeration impeller located inside a casing that is open at the bottom. The casing has flow-directing vanes that form the stator. An air-draft tube or holes for air intake are located in the top plate of the casing along the impeller rim. As an option, an additional mixing turbine can be secured to an extended shaft beneath the aeration turbine. When this device is driven by a motor the aeration impeller pumps liquid in the stator. A vacuum is generated at the impeller rim and ambient air is sucked into the stator through the air intake and is dispersed into fine bubbles. The air-water mixture is discharged from the stator downward into the bulk liquid of the vessel. The impeller-pumping effect causes liquid circulation and establishes mixing patterns in the vessel. Because the stator reduces liquid circulation intensity, the additional mixing turbine is used in deeper vessels.

The Leningrad Institute of Chemical Machine Building[1] has developed a closed-turbine, downdraft aerator (Figure 5.10) and two models, 22 and 40 kW (30 and 53 hp), which are presently manufactured in the USSR. Because these aerators do not produce any droplets or mist,

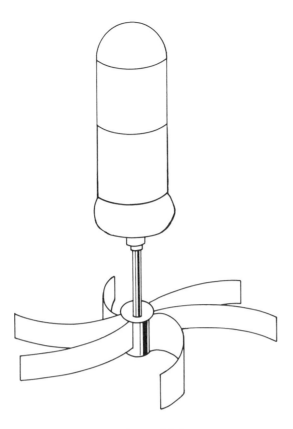

FIGURE 5.7—Axial aerator with open blades.

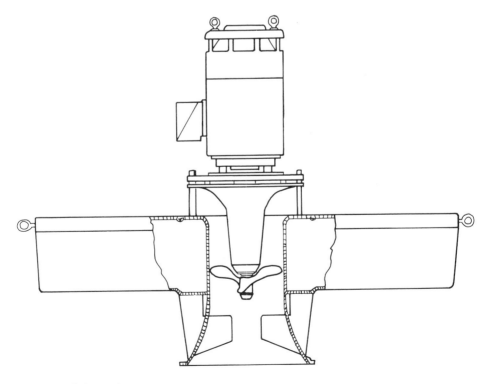

FIGURE 5.8—High-speed aerator.

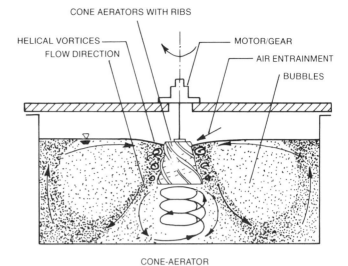

FIGURE 5.9—Open rotor downdraft aerator.

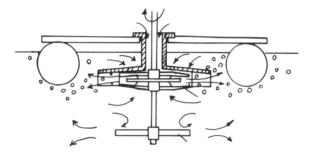

FIGURE 5.10—Closed turbine downdraft aerator, USSR.

they operate successfully at extremely low temperatures in the aerated ponds of several pulp and paper mills located near the Artic Circle.

Forced Air Propeller. Downdraft-tube, vertical-axis aerators with forced air have a propeller-type impeller located in a well which is open at the top and bottom, a motor for driving the propeller, and a flow-through, air-distribution grid. This grid is in the same well either at the well base or immediately below the propeller. Optional devices for uniform-flow distribution across the well section are available. Air delivered from a blower is distributed in bubble form across the well. The propeller produces water downflow, which retains these bubbles in the well. If the downflow water velocity is greater than the rising bubble velocity, the air-distribution grid must be located immediately under the propeller. The bubbles are discharged at the well bottom and rise in the rest of the vessel volume. If the propeller is designed to produce a flow velocity lower than the velocity of rising bubbles, the grid is located at the well base and the bubbles are discharged at the well top. Other modifications are also possible. Liquid circulation and mixing patterns are determined mainly by the propeller-pumping effect.

One type of downdraft-tube aerator with forced air is shown in Figure 5.11. The aerator has a propeller pump located in a draft tube and is driven by a gear-motor assembly, an air sparger that is located under the impeller that is fed with air from a blower. Antivortex intake baffles are located above the draft tube.

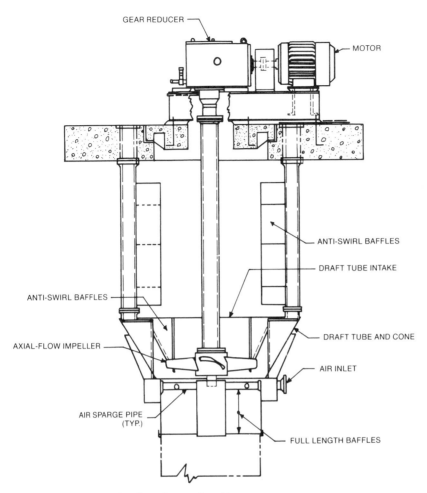

GEAR REDUCER

MOTOR

ANTI-SWIRL BAFFLES

DRAFT TUBE INTAKE

ANTI-SWIRL BAFFLES

DRAFT TUBE AND CONE

AXIAL-FLOW IMPELLER

AIR INLET

AIR SPARGE PIPE
(TYP.)

FULL LENGTH BAFFLES

FIGURE 5.11—Draft tube submerged turbine aerator.

Submerged Mechanical Aerators with Vertical Axis

Submerged mechanical aerators with vertical axis are divided into three groups: turbine, impeller (suction), and porous-rotating discs.

TURBINE AERATORS. Turbine aerators can be provided with mechanical means for drafting air in the liquid or with spargers that are fed with the air by blowers.

Turbine Aerators with Mechanical Air Draft. These aerators have surface and submerged turbines secured on the same vertical shaft. The upper turbine has air-intake holes located near the shaft. When the aerator is activated, an intensive vortex forms under the upper turbine and along the vertical shaft. Air sucked from the atmosphere through the holes in the upper turbine enters the vortex and forms an air column that propagates to the lower turbine. The lower turbine breaks the air into bubbles and disperses them in the bulk liquid volume of the aerated vessel. Additional aeration occurs at the water surface in a similar fashion to surface aerators with vertical axis. The flow patterns are determined by three major components.

1. Vertical circulation involves upflow sections under the upper and lower turbines, a downflow section over the lower turbine, horizontal flows from the turbines toward the vessel walls, and from the walls toward the turbines at the bottom and the mid-elevation between turbines.
2. The water mass rotating in the direction of the aerator rotation.
3. Vessel shapes determine flows.

Turbine Aerators with Forced Air. Turbine aerators with forced (sparged) air (Figure 5.12), also called combined diffused air-mechanical aerators, have a submerged, open-bladed turbine on a vertical shaft driven by a gear-motor assembly, with an air sparger located under the turbine. Usually, one or more additional turbines are secure to the same shaft above the bottom turbine.

Aeration is performed by sparging air under the rotating bottom turbine. This produces high turbulence, breaks air into bubbles, and disperses them in the bulk liquid in the vessel. Upper turbines provide additional mixing. Liquid circulation patterns are determined by the three major flow components; they are similar to those described for aerators with mechanical air draft.

One example of this type of turbine aerator is powered by a standard direct connected gear-motor drive. Both motor and drive are secured to a beam structure that spans the aeration tank. The shaft with impellers is bearing supported in the gear-reducing, drive head. Shaft alignment is assured by a steady bearing. Two or more turbine-type impellers are used; the bottom impeller is located near the tank bottom above a sparge ring, the upper impeller is submerged 0.75 m (30 in.) below the water surface. The sparge ring is designed to emit air at the periphery of the lower impeller. Aerator efficiency varies depending on impeller size and position.

Two modifications of an aerator include a radial-flow impeller with stabilizing ring (Figure 5.13) and an axial-flow impeller (Figure 5.14). Both have a power range between 0.8 and 110 kW (1 to 150 hp). As shown in Figure 5.13, the radial-flow impeller has vertical blades enclosed in a stabilizing ring that is closed at the top. The air is sparged under the ring, is dispersed into bubbles in the liquid flowing through the impeller, and is dispersed in the horizontal (radial) direction. This aerator is especially suitable for shallow basins. The aerator shown in Figure 5.14 has a radial-flow impeller with pitched blades that directs the air-water mixture from the sparge ring downward. This aerator type is more suitable for deeper tanks.

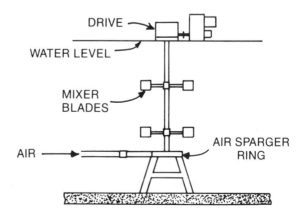

FIGURE 5.12—Turbine aerator with forced (sparged) air.

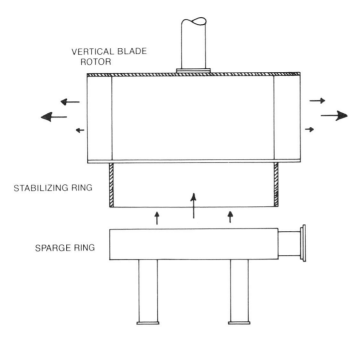

FIGURE 5.13—Radial flow impeller with stabilizing ring.

IMPELLER AERATORS. Impeller aerators can be operated with air suction created by the impeller or assisted by a blower (forced air).

Impeller aerators consist of an impeller enclosed in a casing or a shroud submerged in the liquid to be aerated, and an air-suction pipe. Centrifugal, radial-axial, and axial rotors can be used. The upper end of the air suction pipe originates above the liquid level in the vessel and opens to the atmosphere. The lower end of the air suction pipe is located in the zone of maximum vacuum produced by the rotor. Sometimes the same pipe is used for both air and water suction.

When the aerator is started, the vacuum greater than the water pressure at the suction pipe lower end increases, and sucks ambient air to the rotor that mixes it with water. In aerators with centrifugal impellers, especially those with tooth-shaped vanes, "soft cavitation" occurs, increasing OTE. When impellers are submerged to a substantial depth, the air supply can be increased by using a blower.

The liquid circulation and mixing pattern in vessels with impeller aerators are predominantly determined by an impeller-pumping effect. Centrifugal impellers usually provide lower flows than axial or radial-axial impellers. Centrifugal impellers also provide less mixing, while axial impellers provide a greater mixing intensity.

An impeller aerator consists of an impeller, a stator, and a motor all combined into a submersible unit. An air-intake pipe is connected to the bottom of the unit. The aerator nominal power is between 2.2 and 95 kW (3 to 125 hp). Actual power drawn depends on aerator submergence. The aerator can be operated as a portable, stationary, or mobile unit (self-propelled) with floats. Advantages associated with the use of a submersible unit include the absence of splashing, mist, and icing, low noise, and simplicity of operation.

In the portable model, the aerator is placed directly on the tank floor. Its own weight and low center of gravity ensure high stability. The aerator can be lifted from the tank with a crane hook for maintenance or inspection. This type of installation is particularly useful in large-size earth-constructed tanks and can be operated at various water levels.

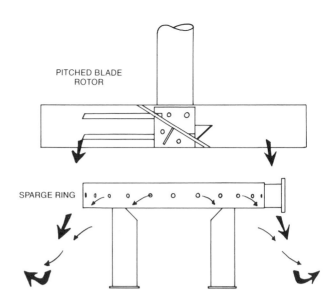

FIGURE 5.14—Axial flow impeller with sparger ring.

The stationary model is intended for concrete tanks. The aerator is lowered between two guide beams anchored to the tank. These beams can also serve as steadying pipes for light semi bridge construction. The unit then can easily be lifted above the water surface for inspection and maintenance.

The mobile-floating model is provided for aeration of lakes, pools, and rivers. The aerator can be mounted under a rotating bridge in a tank or move back and forth on a guideline under its own propulsion.

For deep submergence applications, an air blower can be used to assist the self-aspiration action. Accessories such as a service float, motor controls, or support legs can be provided.

One type of submersible aeration device which has been used frequently for aerating ponds consists of a motor, a hollow-rotating shaft, a unit housing, an impeller and diffuser, a vortex shield, and a hard-coated sleeve. The hollow 1.1 m (44 in.) long shaft is attached to the motor shaft, rotates at 3450 rpm and drives a three bladed-impeller. The impeller accelerates water to a velocity high enough to cause a pressure drop along the liquid surface so that the atmospheric air is drawn through the hollow shaft and broken into bubbles in the liquid flow. The liquid flow discharged by the aerator induces mixing patterns in the vessel.

These aerators can be installed either on booms (e.g., in ponds with several aerators and low-water level fluctuations) or on floats (e.g., when a few units are dispersed over a greater area or the water level substantially fluctuates). The aerator inclination can be adjusted from a vertical to a nearly horizontal position.

The major advantage of this aerator type is that numerous units can be placed in a pond to create a mixing regime that can be induced to provide solids suspension and eliminate dead zones.

POROUS-ROTATING DISCS. Porous-rotating discs with forced air or oxygen have a cored impeller with a porous top plate secured to a hollow shaft. Mixing blades are attached to the impeller. During operation, air or oxygen is blown through the hollow shaft in the impeller core and passes through the porous plate. The bubbles are sheared off the porous plate top by the friction forces between the liquid and rotating disc. The shearing effect ensures micron-sized bubble formation.

The bladed impeller produces vertical circulation and general rotation of the bulk liquid. This is also influenced by vessel geometry. This aerator type has been used in fermentation. As with other porous plate units, filtering of the air or oxygen supply is important.

An example of a porous rotating diffuser is shown in Figure 5.15. It has a 2 m (7 ft) diameter, submerged-rotating disc located approximately 1 m (3 ft) above the vessel floor, and is mounted to a 150 mm (6 in.) diameter hollow shaft. Ceramic diffusion tiles are attached in sections and are mounted into a plastic frame 200 mm (8 in.) wide near the periphery of the disc. These plates form circular diffusion bands on both the upper and lower surfaces, which are adjacent to the tapered disc edge. Mixing impellers are mounted on the disc top and bottom surfaces to provide the shear that aids in unique micron-sized bubble formation, solids mixing, and high-oxygen transfer.

The disc and attached shaft are rotated by a gear reducer and motor that operates at approximately 80 rpm. Oxygen is transferred at approximately 1.7 atm (25 psig) from a supply line to the hollow shaft via a rotating seal; it is then distributed from the shaft to the diffusion assemblies.

As the air or oxygen emerges from the ceramic tile in the diffusion assembly, the disc rotation and the mixed liquor flow shear off the oxygen bubbles to produce fine bubbles.

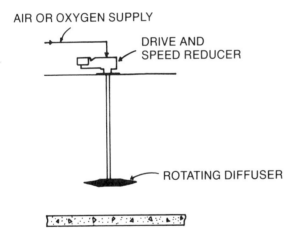

FIGURE 5.15—Submerged porous aerator with forced air.

MECHANICAL AERATORS WITH HORIZONTAL AXIS

Mechanical aerators with horizontal axis are divided into two groups—surface and submerged aerators.

SURFACE AERATORS. Original brush-type aerators, known as Kessener aerators, had a horizontal cylinder rotor with bristles submerged in the bulk liquid of the vessel, approximately to the half-diameter. Now, angle steel, other shape steel, or plastic bars are used instead of bristles. One modification of these aerators makes use of flat or curvilinear steel or plastic blades attached to the impeller.

When the aerator is driven by a motor, the bars or blades drive the air into the water and throw water jets and droplets into the air, which results in oxygen diffusing into the air-water interface. Simultaneously, the liquid is propelled by the rotor, which provides mixing in the vessel.

Enclosures over the top section of the aerator are used to reduce mist and droplet dispersion into the air and to prevent or reduce the freezing problem. In some applications flow-guiding devices are used in the vessel to improve horizontal liquid flow patterns.

Figure 5.16 shows a typical surface aerator for channel aeration. Each of the horizontal blades has wide metal teeth. These blades are supported by end plates. The entire rotor assembly is supported by field-replaceable, stub shafts. Rotor aerators are used in aeration channels to approximately 2 m (6 ft) liquid depth, and are available in several lengths. Impeller blades have been developed to withstand impact from floating debris such as the ice floes expected in severe winter operations. Oxygen output and power drawn are functions of both rotor speed and immersion.

FIGURE 5.16—Surface aerators for channel aeration.

SUBMERGED AERATORS. Various modifications of the submerged horizontal-axis aerators such as the disc and paddle have been developed. One such device, the disc aerator, has been used in numerous installations and is described below.

Disc Aerators. Disc aerators consist of wafer-thin circular plates mounted on a horizontal shaft. The discs are submerged in the water for approximately one-eighth to three-eighths of the diameter and enter the water in a continuous, non-pulsating manner.

One manufacturer's disc aerator is made of a non-corrosive plastic material that relies on recesses and nodules on the face of the disc to both propel the liquid and transfer oxygen. The recesses and nodules introduce entrapped air below the surface as the disc turns in the water. The discs are provided in two half-sections which clamp together a standard circular shaft. Spacing of the discs on the shaft can be variable and depends on the particular oxygen needs of the system and on basin geometry.

Typical power requirements are 0.1 to 0.75 kW/disc (0.15 to 1.00 hp/disc).

*F*ACTORS AFFECTING MECHANICAL AERATION PERFORMANCE

The performance characteristics of aeration systems[2-4] depend on several major groups of factors (Figure 5.17), including geometric and dynamic parameters, and composition and temperature of the process water. With respect to mechanical aeration systems, the geometric characteristics include the type and sizes of aeration device, the shape and sizes of the aeration basin, and the position of the aerator or aerators in the basin. The dynamic characteristic of the system is the rotational speed of the aerator. Geometry of the system and the rotational speed of aerators largely determine the process hydrodynamics that, in general, can be described by the velocities, pressures, and flow depths in the system. To some extent, the process hydrodynamics depend on the physiochemical properties of the process water—primarily viscosity, surface tension, average solids concentration, particle size distribution, and the distribution of the solid mass over the aertion basin volume. The hydrodynamics of the system and the physiochemical properties of the solution determine the performance characteristics of aeration systems. From a practical standpoint, the important characteristics include the OTR, liquid mixing, and power demand. The OTR depends on the values of the mass transfer coefficient and the oxygen equilibrium concentration. Two aspects of the liquid mixing should be considered: the velocities of flow that provide particle suspension (usually expressed as a minimum bottom velocity), and the hydraulic residence time distribution that is often expressed through the coefficient of longitudinal dispersion.

Numerous experimental studies[5-20] have been devoted to evaluating fundamental parameters and major groups of factors illustrated in Figure 5.17. Such studies usually are related to a specific type of aerator and they cover only a limited set of the systems variables. This section briefly summarizes current understanding of factors affecting the performance of mechanical aerators.

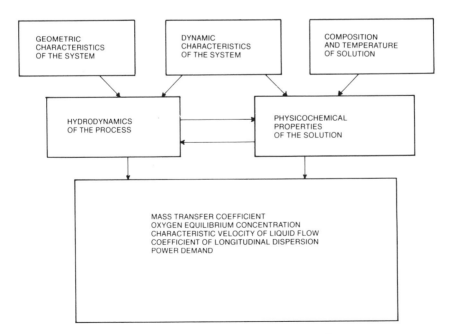

FIGURE 5.17—Major parameter groups that affect aerator performance.

CHARACTERISTIC CURVES. The most important characteristic curves of a mechanical aerator are presented by (a) the total oxygen transfer rate (OTR) at standard conditions (mass of O_2 per hour), (b) the shaft-power demand (W or hp/hr), and (c) the aeration efficiency of the aerator (mass of O_2 per kW-hr). For a given aerator, such curves may be conveniently plotted against one or another operational parameter.

For a surface-turbine aerator, such curves usually are presented as a function of the aerator submergence. Generally, the top level of the blade is used as a reference point for submergence. A typical set of characteristic curves (Figure 5.18)[13] demonstrate that three operational regimes of the aerator operation are possible: spraying, bubble-trapping aeration, and mixing without significant aeration. Spraying regime occurs when the top of the blade is not submerged in water ($H_c < 0$). A low-power demand and a low-pumping rate are typical for this regime. Although the aeration efficiency in this regime is high, the total oxygen transfer rate from droplets is low. Accordingly, this aeration regime is not practicable. A bubble-trapping aeration regime occurs at a moderate submergence of the aerator rotor. This regime is characterized by the formation of a circular hydraulic jump around the impeller. At the optimum submergence, the aeration efficiency reaches the maximum. In Figure 5.18, optimum submergence is approximately at H_c equal to 4 cm. Under the optimal conditions, the hydraulic jump is located at the rim of the impeller. At a submergence greater than optimal, the position of the hydraulic jump gradually shifts on to the rotor so that

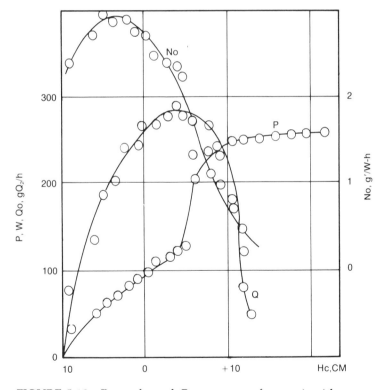

FIGURE 5.18—Power demand, P, oxygen transfer capacity, (shown here as Q), and oxygen transfer rate, N_o, as function of the aerator submergence. (Note: Low speed aerator, surface type, with Diameter = 15 cm, Rotational speed = 450 rpm, h = 2 cm, b = 7 cm, and n = 6. H_c = O when top of blade is at the water surface; H_c > O when top of blade is submerged.)

the rotor edges become submerged. This causes a rapid increase in the power demand and a significant reduction in the quantity of air trapped by the jump. Simultaneously, the flow rate of liquid increases, thus producing a somewhat greater mixing effect in the tank. The total oxygen transfer rate diminishes simultaneously with the increasing power draw. Consequently, this operational regime is also not practicable.

The operational regime also depends on the aerator rotational speed. For a given submergence, the spraying regime occurs more readily at greater speeds, and lower speeds favor the mixing regime.

For other types of aerators (e.g., rotor aerators), the characteristic curves are quite similar to those shown in Figure 5.18. For aerator types such as submerged and vertical-axis aerators, the characteristics functions depend on the air-flow rate, the aerator speed, and the aerator submergence. Shapes of characteristic curves for these aerator types may be different; however, the curves can be used to identify the domain of the system's optimal parameters (i.e., the domain in which the specific efficiency of the aerator and the total OTR are sufficiently high). For the same domain, the mixing characterics (i.e., the characteristics velocity of liquid flow, dispersion coefficients, and parameters of the homogeneity of the oxygen and biomass distribution in the aerated volume) are also determined. The oxygen transfer and mixing characteristics of an aerator can be used either for scaling up the aeration equipment, or for the design of the oxygen transfer equipment for wastewater-treatment processes.

Scaling data are often presented in the form of one or another correlation between the geometric and dynamic parameters of the aeration system. Correlations are often developed by the use of the dimensional analysis. For example, Horvath[9], Schmidtke[21, 22], Khudenko and Shpirt[12] and many others developed several such correlations for surface-turbines, bladed cage-impellers and other aerators. The accuracy of some of these correlations is claimed to be plus-or-minus 20%. These correlations, however, describe only a near-optimum domain. Moreover, small changes in certain process conditions (e.g., the aerator submergence as shown in Figure 5.18) may result in a drastic change in the power demand and a substantial change in the specific efficiency of the aerator. These dramatic changes are not always accurately described by the power type correlations such as those commonly used in dimensioal analysis. Accordingly, Stukenberg[18] recommends obtaining the characteristic data either from shop or field testing of aerators.

PERFORMANCE OF MECHANICAL AERATORS UNDER STANDARD CONDITIONS

There are numerous literature sources[2–4, 11, 12, 14–16, 23] that provide data on the aeration efficiencies of mechanical aerators. Usually, the aeration efficiency is given for standard conditions (e.g., clean water, 20°C, normal atmospheric pressure), assuming complete-mix conditions. The typical ranges of aeration efficiencies of aeration for various aerator types are given in Table 5.1.

The data presented in Table 5.1 must only be used to estimate the energy requirements and preliminary designs of aeration systems. The final design should be based on the shop- or field-test data and should consider the specifics of application.

OTE IN PROCESS WATER

The oxygen transfer in process water differs from that at standard conditions because of a lower driving force and a lower mass transfer coefficient. Additionally, the interfacial area in the process water may also either increase or decrease transfer efficiencies. There are no reliable data on this matter. Accordingly, the existing design procedures consider changes only in the driving force and in the mass transfer coefficient, $K_L a$.

The driving force of the oxygen transfer in the process water is the difference between the DO saturation concentration and the actual concentration of the DO in the bulk liquid: $(\beta C^*_T - C)$, where subscript T signifies the process temperature and the value of β reflects the reduction in the saturation concentration caused by the admixtures to the water. C^*_T is the spatially averaged DO saturation concentration at temperature T in clean water. The mass transfer coefficient, $K_L a$ in the process water is corrected for the temperature and the effect of admixtures as described in Chapter 3, where $\propto$ and θ are the alpha factor and the temperature correction coefficient, respectively.

By making the process-water corrections to the values of the driving force and the mass transfer coefficient, the expression for the specific OTR can be presented as:

$$N = N_o \frac{\propto (\beta C^*_T - C)}{C^*_T} \cdot \theta^{T-20} \qquad (2)$$

Where:

N$_o$ and N were defined earlier in this chapter.

The value of β is usually assumed to be 0.92 to 0.98. The θ value is commonly accepted to be equal to 1.024. The value of $\propto$ varies with the water quality, aerator type, and the operational regime of the system. Only approximate data exists on the $\propto$ values. Typical available information is summarized in Tables 5.2 and 5.3.[4, 20, 25–30]

It should be emphasized that it is often necessary to obtain $\propto$ by pilot and full-scale aerator testing under process conditions. The data presented here are mainly intended to illustrate the range of the factor variability rather than recommend certain design values.

Table 5.1. Typical ranges of specific aeration efficiency for various aerator types.

Aerator Type	Aeration Efficiency + $k_g O_2/kW \cdot h$	
	Standard	**Field**
Surface centrifugal (low speed)*	1.2–3.0	0.7–1.4
Surface centrifugal wiht draft tube	1.2–2.8	0.7–1.3
Surface axial (high speed)*	1.2–2.2	0.7–1.2
Downdraft open turbine	1.2–2.4	0.6–1.2
Downdraft closed turbine	1.2–2.4	0.7–1.3
Submerged turbine, sparger	1.2–2.0	0.7–1.1
Submerged impeller	1.2–2.4	0.7–1.1
Surface brush and blade	0.9–2.2	0.5–1.1

*Approximate; rpm unknown

+ Design values of Aeration Efficiency for clean water conditions and corrective factors for field conditions should be obtained from equipment manufacturers and from tests on the particular wastewater. If specific data are not available, conservative transfer rates should be used.

Table 5.2. Typical Values of ∝-Factor* for selected wastewater types

Wastewater Type	BOD$_5$ (mg/L)		∝-Factor	
	Influent	Effluent	Influent	Effluent
Pulp and paper	187	50	0.68	0.77
Kraft paper (mixed wastewater)	150–300	37–48	0.48–0.68	0.7–1.11
Bleached paper	250	30	0.83–1.98	0.86–1.0
Pharmaceutical plant	4500	380	1.65–2.15	0.75–0.83
Synthetic fiber plant	5400	585	1.88–3.25	1.04–2.65
Municipal wastewater	180	3	0.82	0.98

*Low-speed, surface aerators. There is no consensus on methods for ∝ determination. Recent research suggests that ∝ values may be lower and more variable than the values listed in the table.

Table 5.3. Typical Values of ∝-Factor* for selected aerator types.

Aerator Type	Tank or Plant Size	Process Conditions	Test Method	∝ Factor
Simplex	1000 m^3	ASP	N/A	>1.1
Simplex cone	Full scale	ASP	N/A	1.1
Surface	Full scale	ASP	N/A	0.93
Cone	Full scale	ASP	SS, NSS	1
Rotor	Full scale	N/A	SS, NSS	1
Draft tube	5700 m^3/d	OD, F:M = 0.14	Off-gas	0.66–0.77
Draft tube	5100 m^3/d	ASP, PACT	Off-gas	0.3–0.6
Draft tube	2800 m^3/d	OD, F:M = .086	Off-gas	0.81–0.87
Disc (Orbal)	680 m^3/d	OD	SS, NSS	0.95–1.05
Disc (Orbal)	730 m^3/d	OD, refinery	SS, NSS	0.98–1.03

SS   = steady state, clean water
NSS = nonsteady state, process water
ASP = activated sludge process
OD   = oxidation ditch
PACT = Powdered activated carbon treatment
N/A = Not available
* Municipal wastewater, unless otherwise stated. There is no consensus on methods for ∝ determination. Recent research suggests that ∝ values may be lower and more variable than the values listed in the table.

MIXING CHARACTERISTICS OF MECHANICAL AERATORS

With respect to solids suspension and oxygen dispersion throughout the bulk liquid volume, the two criteria commonly used are (a) the minimum bottom velocity of flow, and (b) a minimum power input per unit volume.

Using one or another criterion depends mainly on the type of aeration facility, and also on the preference of the researcher conducting the study. There is general agreement that for oxidation ditches the bottom velocity is the preferred criterion. For aeration tanks and aerated ponds and lagoons, either the bottom velocity or the power input is used.[17,24] The minimum velocity of flow in oxidation ditches is usually assumed to be 0.3 to 0.6 m/s (1 to 2 ft/s). The minimum power input per unit volume is usually specified by the aerator manufacturer. Sometimes, equivalent data are presented, such as the tank volume that may be serviced by a single aerator. Typical data for an activated sludge plant with conventional-loading rates are given in previous sections. In aerated lagoons, Rich[31] suggests that the power requirements for solids suspensions start at 1.6 W/m³ (8 hp/10⁶ U.S. gal) but that to achieve practically complete-mix conditions, power requirements should be from 4 to 6 W/m³ (20 to 30 hp/10⁶ U.S. gal).

CALCULATION OF THE NUMBER OF AERATORS

Assuming steady-state conditions and a completely mixed system, and considering Equation 2, the following formula can be used to calculate the number of aerators for the entire plant:

$$m = \frac{R \cdot Q}{N \cdot P} \qquad (3)$$

$$= \frac{R \cdot Q \cdot C_T{}^*}{\propto N_o \, P \, (\beta C_T{}^* - C) \, \theta^{T-20}}$$

Where:

m = number of aerators
R = oxygen uptake rate, kg O_2/m³,
Q = flow rate, m³/h,
N = field aeration efficiency, kg O_2/kW • h, and
P = power requirement for a single unit, kW.

The power requirements suggested by the manufacturer should be checked against the projected power input per unit volume mP/V, where V is equal to the total volume of aeration tanks.

APPLICATIONS OF VARIOUS TYPES OF MECHANICAL AERATORS

Selection and specifying mechanical aerators is discussed in several sources.[1–7, 13–20, 24–26, 32–36] The major factors that need to be considered in the process are (a) type of treatment facility (e.g., conventional, high rate, extended aeration, activated sludge process, oxidation ditch, and aerated lagoon); (b) plant capacity; (c) variability in the flow rate and the concentration of wastewater; and (d) climatic conditions.

Low-speed, surface-turbine aerators (i.e., centrifugal impellers) were first used predominantly in complete-mix aeration tanks at conventional- or high-loading rates. Later, these aerators also were applied to low rate processes such as extended aeration. The plant capacities were usually not greater than 50,000 m³/d (13 U.S. mgd). Current practice calls for aeration facilities to have several aeration tanks operated in parallel. The typical depth of the tanks is 4.5 m (15 ft); the tank diameter, or the length of the wall of the square tank, generally does not exceed 5 to 7 diameters of the aerator rotor. Such sizes provide complete-mix conditions in the tank. A central-draft tube additionally improves the solids suspension. Because low-speed aerators have bulky reducers, they are heavy and are usually mounted on fixed supports.

Under variable-flow rates and organic loadings, the OTR of low-speed, surface aerators is controlled by the use of either variable- or dual-speed motors or variable-aerator submergence. The aerator submergence is commonly controlled by the use of level controllers at the tank outlet.

In cold climates, aerators are often provided with a shroud that prevents the freezing of mist and droplets on the adjacent structures, wire lines, and rods. The shrouds reduce the OTR. Winter conditions, however, are usually not critical from this standpoint, thus ensuring that the oxygen supply remains adequate.

In recent years, low-speed, surface aerators have also been used to simultaneously aerate and pump the return sludge from the secondary clarifiers (Figure 5.19), for circulation of the liquid in ring- or oval-shaped treatment units (Figure 5.20), and for aeration of low-rate, biological and lagoons.[12, 17]

High-speed (radial-axial and axial) surface aerators were first used in aerated ponds and lagoons. Presently, this type of aerator is also used in various modifications of the activated sludge process. The plant capacities may reach 50,000 m³/d (13 U.S. mgd). The aeration tank shape may be either circular or square. Tank size determination should be based on the aerator manufacturers' specifications and the shop and field testing.

Because high-speed aerators are of light weight, they are often mounted on floats. This is especially convenient for lagoon applications where the water level may change substantially. Wave action is also a factor and should be considered.

High-speed aerators are also portable, and have low initial and installation costs which makes it relatively easy to add to an existing system. Their disadvantages include a lower aeration efficiency and poor mixing capability compared to low-speed, surface aerators.

In cold climates, high-speed aerators produce greater cooling of the aerated liquid than the low-speed aerators, thus reducing biological activity. Icing of the floating unit may occasionally result in capsizing; therefore, flotation stability should be a consideration.

Closed-turbine downdraft-surface aerators may have a slightly lower aeration efficiency than high-speed aerators. They are, however, ideally suited for winter operation. Such aerators are usually installed on a fixed mounting in either a lagoon or an aeration tank. The control of the oxygen transfer in lagoons is usually based on turn on-turn off operation of one or several units. In aeration tanks, oxygen transfer is controlled by changing the water level in the tank. The downdraft aerators with a second turbine on the extended shaft can be used in reservoirs of 6 m (20 ft) depth and greater.

The many modifications of downdraft aerators with forced air are used for various applications. An aerator with a U-tube is used in oxidation ditches, but a vertical-downdraft tube is used in activated sludge process aeration tanks. These aerators usually have lower aeration efficiency than do low-speed surface aerators. They are, however, very reliable for winter operations and they afford a good control of oxygen transfer by changing the flow rate of compressed air. A disadvantage of this aerator type is that air-blowing equipment is required.

Submerged mechanical aerators with vertical-axis and surface turbines are used for larger activated sludge plants, having a throughput of

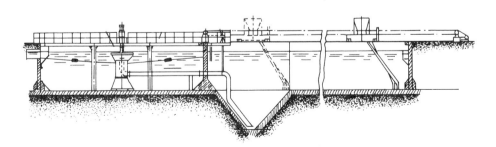

FIGURE 5.19–Mechanical surface aerator with draft tube (also used for sludge recycle).

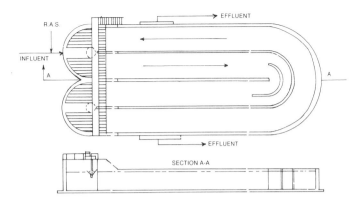

FIGURE 5.20—The "Carrousel" layout—a ring- or oval-shaped treatment unit.

up to 600,000 m³/d (160 U.S. mgd). Such aerators can be installed in deep tanks, usually 6 m (20 ft) or deeper. Because of an additional power draft required by the lower turbine, the aeration efficiency of these aerators is less than that of the low-speed, surface turbines. Mixing capacity, however, is superior.

Submerged mechanical aerators of the impeller (suction) type have a power range suitable for small (10,000 m³/d or 2.5 U.S. mgd) to medium-sized plants (50,000 m³/d). These aerators are very convenient for cold-weather operation. Aeration efficiency and mixing capabilities, however, are lower than those of low-speed, surface turbines. The use of compressed air instead of the natural draft in these aerators provides additional control of the aeration efficiency.

Turbine aerators with sparged air are normally used at medium-sized (50,000 m³/d) and larger plants. These aerators may provide a somewhat lower aeration efficiency when compared to other aerators. They provide the largest power input per unit weight of aeration device, however, in the category of large aerator. They also have good mixing capabilities and can be installed in deep tanks with liquid depth of 6 m or deeper. The controllable flow of air affords a relatively wide range of aeration efficiency control. The aerators are convenient for winter operation and are used, in fact, in several cities and at pulp and paper plants in the Arctic and Siberia.

Various modifications of rotor aerators and other aerators with horizontal axis are used at small-capacity treatment plants with oxidation ditches or similar process modifications. The aeration efficiency of these aerators is similar to slow speed turbines. These aerators, however, are capable of propelling the liquid in the aeration channel at a substantial velocity, 0.3 to 0.6 m/s (1 to 2 ft/s). In cold climates, the winter operation presents a problem because of mist and freezing droplets. This may be partially alleviated by the use of aerator enclosures.

Combined diffused–air mechanical aerators with horizontal axis reportedly provide superior aeration efficiency and superior liquid propulsion in oxidation ditches. Oxygen transfer can be controlled by using the variable air flow. Because neither water droplets nor mist are generated, this type of system should be appropriate for cold climates.

*O*PERATION AND MAINTENANCE CONSIDERATIONS

The effective operation of aeration equipment can minimize power consumption and maximize performance. Operation and maintenance

---

should focus on (a) control of the preset oxygen concentration in the liquid being aerated, (b) provision of at least the minimum-required mixing intensity, and (c) inspection and service of the aeration equipment to provide an uninterrupted operation.

Usually, the preset oxygen concentration in the liquid under aeration is chosen to be 1.0 to 2 mg/L. At usual temperatures (about 20°C or 68°F), the power consumption for maintaining 4 mg/L of DO instead of 2 mg/L would grow by 35 to 40%. Because the power consumption for aeration is one of the major operational expenses, the need for oxygen concentration control in the bulk liquid cannot be overemphasized. The control technique depends on the aeration equipment. Control techniques such as listed in Table 5.4 can be used.

Speed control is not usually used at small plants because the cost of large-size variable-speed gear motors is very high, and because the efficiency of such units is low. Changes in the submergence can be provided more readily by the use of controllable outlet weirs. The on and off control should be used, within the limitations of the allowable number of such operations per hour (usually 4 to 8).

The mixing capability of the aeration system should be evaluated before system installation. This should be based on the manufacturers' data as well as shop and field testing. In some cases baffling may be required to improve mixing and reduce hydraulic surges which can cause gear box failure. During routine operation, the minimum mixing intensity that provides solids suspension and oxygen supply throughout the tank liquid volume should be determined and maintained thereafter.

Maintenance of mechanical aerators involves the following:

1. periodic preventive service such as oiling and greasing of bearings and gears, tightening of fastening devices such as bolts and nuts, checking floats, and checking electric cables;
2. logging data on power usage by collecting data on energy use, aerator submergence, and air flow rate, if applicable;

Table 5.4. Techniques for controlling oxygen concentration in aeration facilities.

	Applicable Control Technique			
Aerator Type	Submergence	Speed	Air Flow	On-Off
Surface centrifugal (Low speed)	Yes	Yes	No	Yes
Surface axial (High speed)	No	No	No	Yes
Downdraft open turbine	Possible	Possible	No	Yes
Downdraft closed turbine	Possible	Possible	No	Yes
Downdraft closed turbine with forced air	No	No	Yes	Yes
Submerged sparged turbine	No	No	Yes	No
Submerged impeller	No	No	No	Yes
Surface rotor and disc	Yes	Yes	No	Yes
Horizontal axis air-mechanical	No	No	Yes	Yes

3. logging data on time-in-use for all aeration units and specifying the time used for repairs; and
4. analysis of equipment failures and interruptions in operation.

For combined mechanical-diffused air aerators, the operation and maintenance of the blowing and air-distribution equipment should be provided in a similar fashion to that described in Chapter 4.

REFERENCES

1. Hovis, J.S. and McKeown, J.J., "New Directions in Aerator Evaluation." Seminar Workshop on Aeration System Design. Testing, Operation, and Control, W.C. Boyle (Ed.), U.S. EPA Cincinnati, Ohio (August, 1982).
2. Knop, E., Bischofsberger, W. and Stalmann, V., "Versuche mit Verschiedenen Beluftungssystem in technischen Mabstab. Vulkan Verlag" Essen, W. Germany (1964).
3. Knop, E. and Kalbskopf, K.H., "Energetische und Hydraulische Untersuchungen an mechanischen Beluftungssystemen" *Das Gas und Wasserfach*, **10,** 110 (1969).
4. Kayser, R., "Experiences with Different Aeration Systems in Germany." Seminar Workshop on Aeration System Design, Testing, Operation and Control, W.C. Boyle (Ed.), U.S. EPA, Cincinnati, Ohio (August, 1982).
5. "Air Diffusion in Sewage Works." *J. Water Pollut. Control Fed.*, **33,** (1950).
6. Berk, W.L., "Increased Loading for Existing Lagoons."; "Why Settle for Only Secondary Treatment?"; and, "Treat Your Tannery Waste to an Oxidation Ditch." undated, Lakeside Equipment Corp. Bulletin, Bartlett, Ill.
7. "Workshop Toward an Oxygen Transfer Standard." *Proc. Asilomar Conference Grounds*, W.C. Boyle (Ed.) U.S. EPA, Washington, D.C. (1978).
8. Harremoes, P., "Dimensionless Analysis of Circulation, Mixing and Oxygenation in Aeration Tanks." *Prog. Water Tech.*, **11,** 3 (1979).
9. Horvath, A., "Modeling of Oxygen Transfer Processes in Aeration Tanks." The 3rd International Conference on Water Pollution Research, Munich, West Germany (1966).
10. Horvath, I., "Some Questions of the Scale-Up Aeration System." *Prog. Water Technol.* **11,** (1979).
11. Johnson, R.L.and Usinovicz, P.J., "Rotor Aerator Field Test." *Proc. Am. Soc. Civ. Eng.*, Env. Eng. Div., Amer. Soc. Civil Eng., New York, N.Y. (1979).
12. Khudenko, B.M. and Shpirt, E., "Hydrodynamic Parameters of Diffused Air Systems." *Water Research*, **20,** 7 (1986).
13. Khudenko, B.M. and Shpirt, E.A., "Wastewater Aeration." Strayizdat Publishing House, Moscow, Soviet Union (1973).
14. McKeown, J.J., "Selected Experience with Aerators Used in the Treatment of Paper Industry Wastewaters." Seminar Workshop on Aeration System Design Testing, Operation and Control, W.C. Boyle (Ed.), U.S. EPA, Cincinnati, Ohio (August, 1982).
15. Norcross, K. and Shell, G., "Application of a Unique Aeration Devise to a Textile Wastewater. *Proc. Ind. Waste Conf.*, *Purdue Univ. Ext. Ser.*, Purdue, Ind. (1979).
16. Opincar, V.I. and Quigley, J.T., "Application of Innovative Technology to Aerated Lagoon Systems." *Calif. Water Pollut. Control Assoc. Bulletin*, **16,** (1980).
17. Smith, G.W., "Oxidation Ditch Aeration Systems—Types and Characterics." Seminar Workshop on Aeration System Design,

Testing, Operation and Control, W.C. Boyle (Ed.), U.S. EPA, Cincinnati, Ohio (August, 1982).

18. Stukenberg, J.R. and McKinney, R., "Experiences in Evaluating and Specifying Aeration Equipment." *J. Water Pollut. Control Fed.*, 1, 66, 49 (1977).

19. Quigley, J.T., "Lagoon Aerator Keeps Solids Suspended Under Heavy Ice." *Water and Sew. Works* (1978).

20. von der Emde, W., "Aeration Developments in Europe." In *Advances in Water Quality Improvement*, Gloyna, E.F. and Eckenfelder, Jr., W.W. (Eds.), Univ. of Texas Press, Austin, TX, 237 (1968).

21. Schmidtke, N.W., "Aeration System Scale Up." Seminar Workshop on Aertion System Design, Testing, Operation and Control, W.C. Boyle (Ed.), U.S. EPA, Cincinnati, Ohio (August, 1982).

22. Schmidtke, N.W. and Horvath, I., "Scale-Up Methodology for Surface Aerated Reactors." *Prog. Water Technol.* 9, (1977).

23. Nestman, F., "Optimization of Mechanical Aerators in a Hydraulic Two-Phase Flow Model and the Scale-Up of the Results for the Prototype." *Proc. 1st Inter. Workshop Scale-Up of Water and Wastewater Treatment Processes*, (1984).

24. Rooke, T.D., "Mechanical Aeration Systems—Types and Characteristics." Seminar Workshop on Aeration System Design, Testing, Operation and Control, W.C. Boyle (Ed.), U.S. EPA, Cincinnati, Ohio (August, 1982).

25. Downing, A.L., Bayley, R.W., and Boon, A.G., "The Performance of Mechanical Aerators." *J. Inst. of Sewage Purification*, 3, (1960).

26. Boon, A.G., "Measurement of Aerator Performance." *Symp. on the Profitable Aeration of Wastewater, London, April 25, 1980*, BHRA Fluid Engineering, Cranefield, Bedford, UK, 13 (1980).

27. Huibregtse, G.L. and Doyle, M.L., "Full Scale Alpha Determination for Orbal Aeration System." Rexnord, Inc., Waukegan, WI, Test Center Report (January, 1982).

28. Applegate, C.S. and Huibregtse, G.L., "Orbal Performance Tests—comparison of Two Types of Discs." Rexnord, Inc., Waukegan, WI, Test Center Report (July, 1978).

29. "Adaptation and Evaluation of existing Methods for Estimating the Oxgyen Transfer Performance of Total Barrier Oxidation Ditches Equipped with Draft Tube Turbine Aerators." U.S. EPA 68-03-1818, Cincinnati, Ohio (1988).

30. "Performance Evaluation of the Fred Hervey Water Reclamation Plant, El Paso, Texas." Parkhill, Smith and Cooper, Inc., El Paso, Tex. (October, 1986).

31. Rich, L.G., "How Maintenance Mechanically Simple Wastewater Treatment Systems." McGraw-Hill Books, New York, N.Y. (1980).

32. "Aeration in Wastewater Treatment." Manual of Practice No. 5, Water Pollut. Control Fed., Alexandria, VA (1971).

33. Campbell, H.J., "Oxygen Transfer Testing Under Process Conditions." Seminar Workshop on Aeration System Design, Testing, Operation and Control, W.C. Boyle (Ed.), U.S. EPA, Cincinnati, Ohio (August 2–4, 1982).

34. Wallace, N., "Lake City Marina Defeats Algae, Weeds, and Ice." *Marina Management/Marketing*, Annual Buyer's Guide, 5,

35. Baars, J.K., "The Effect of Detergents on Aeration: A Photographic Approach." *J. Proc. Inst. Sewage Purification*, 358 (1955).

36. Kormanik, R.A., *et al.*, "The Influence of Tank Geometry on the Oxygen Transfer Capabilities of Mechanical Surface Aeration." Proc. 28th Ind. Waste Conf., 638, Purdue Univ. Lafayette, IN (1973).

Chapter 6
Aeration Control

INTRODUCTION

Over the last 15 years dissolved oxygen (DO) control has been practiced in various ways at many activated sludge treatment plants. During this period, the development of reliable DO probes has made automatic DO control possible.

Manual control is the most commonly practiced DO control method; an operator increases or decreases aeration rate by manually changing either the number of aerators in service, power consumption of mechanical aerators, or the air-flow rate to diffuser. Manual control, which is an important method of DO control, is not the primary subject of this chapter. It is assumed that DO control refers to automatic DO control (i.e., a computer or analog controller adjusts aeration rate according to some predefined control law).

BENEFITS OF DO CONTROL

Although DO control has many benefits, most systems are justified on the basis of energy savings. DO control systems can reduce aeration energy costs by 10 to 58%,[1,2] depending on the wastewater characteristics, and the plant and aeration system design. The savings result from an increased driving force as described in the equations in Chapter 3.

If the average activated sludge DO value is reduced, the driving force increases, thus providing greater oxygen transfer. If the value of DO saturation is 9.0 mg/L, and the average DO value can be reduced from 3.0 to 1.5 mg/L, then the oxygen tranfer rate increases by approximately 25%. Successful control system design will translate this savings into energy savings. Many plants profit from DO control through this mechanism. At plants with manual DO control, an operator often must run the unit process at high mixed liquor DO concentrations to have sufficient DO at high-loading periods. This practice, of course, can cause excessive DO during low-loading periods. An automatic DO control system can avoid much of this waste, and can improve process performance by ensuring that sufficient DO exists at high loading. DO control can only achieve savings if the layout of aeration equipment is such that the rate of DO supply is equal to the rate of demand, particularly in a plug-flow system.

At times of high-oxygen demand, there must be sufficient aeration capacity in each zone of an aeration tank to satisfy the demand without wasting energy by operating beyond the limit of the equipment. At times of low-load the energy input must be sufficient to keep activated sludge in suspension and to avoid wasting energy.

A second advantage of DO control systems is improved process performance. Activated sludge process efficiency is independent of DO concentration above some critical minimum value, which is between 0.5 and 3.0 mg/L, depending on process conditions, operator experience, and need to nitrify.[3] In practice, many plants operate below this minimum during periods of the day because operators are unable to adjust the aeration rate quickly to match changes in plant loading. Also, some plants prevent nitrification by maintaining low DO, and other plants use low DO as a selector or material removal mechanism. There are numerous references to improved plant performance because of DO control. Roesler[8] was among the first to note the improvement. He conducted a plant-scale study at the Renton, Washington, plant, a 24 mgd plug-flow, activated sludge plant that uses diffused aeration. Automatic DO control improved average or mean chemical oxygen demand (COD) removal efficiency, although control did not increase maximum COD removal efficiency. DO control reduced the percentage of time the plant operated poorly (when it removed less than 60% COD). Mean effluent BOD_5 decreased from 12 to 4 mg/L with DO control. Sludge Volume Index also dramatically improved with automatic control.

Jenkins *et al.*[5] have also noted improved plant performance with DO control. They indicate that DO control can avoid extended incidences of low DO, which can cause filamentous sludge bulking, and they cite several case studies where this occurred. They also indicate a relationship between the required DO and the process-loading rate, as measured by F/M ratio. Roesler[4] also implicated DO in poor sludge-settling characteristics, which are caused by excessively high DO concentrations that created dispersed growth. Both extremes can be avoided with DO control systems.

CONTROL SYSTEMS

There is a large body of literature on control systems that covers the range from classical methods to modern control techniques, with applications from servo systems to missiles. Control theory is not a topic for this manual of practice but some background is useful. Control system principles are available elsewhere.[6]

EXAMPLE CONTROL SYSTEMS

Control systems can be divided into two types: feedback and feedforward. Feedback control systems detect the presence of a disturbance in a controlled variable (for example DO concentration), and perform some corrective action by changing the manipulated variable (for example blower speed, number of aerators in service). Therefore, a perturbation in the controlled variable must occur before any control action can be taken. Consequently, feedback control systems are inherently imperfect.

Feedforward control systems sense the disturbance directly and perform a corrective action. Feedforward control systems can theoretically perform perfectly, although in practice this never happens. Feedforward controllers can respond more quickly to disturbances and can minimize deviations from set points. Most feedforward control systems are coupled with a feedback control system. A feedback system is necessary because it is not possible to know precisely what the process response will be to a disturbance. For example, although a control system knows in advance that the influent-flow rate will double at a certain time, it does not know precisely how, nor can it modify process conditions to exactly compensate for the flow disturbance. DO control systems usut-putally do not require such extremely fast or accurate control.

Most DO control systems are the feedback type, although there are some examples of feedforward DO control systems. Examples of feedback, feedforward, and feedback-feedforward control systems are shown in Figures 6.1, 6.2 and 6.3. The control diagram shows a process as a rectangular block and "adders and subtracters" as circles. Lines express signals as information.

Figure 6.1 shows an uncontrolled process. The input signal is shown at the left. The input represents the desired condition or state of the system, or in the case of an aeration basin, it represents the set-point DO concentration. As shown in the figure for the uncontrolled case, there is no information flow from the output back to the input. This condition is equivalent to an operator adjusting aeration rate without ever knowing the DO concentration.

Figure 6.2 shows a feedback control system. The output signal (i.e., measured DO concentration) is fed back to a device that compares it to the set point signal. The difference between the input or desired signal

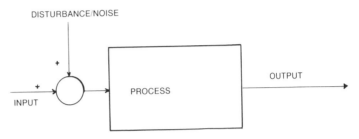

FIGURE 6.1—Uncontrolled process.

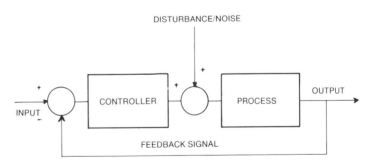

FIGURE 6.2—Controlled process.

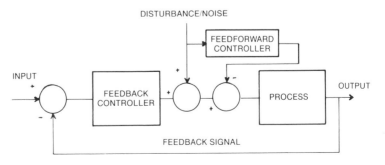

FIGURE 6.3—Feedback/feedforward controller.

and the output signal is fed to a controller, which takes a corrective action, such as increasing the blower speed or opening the inlet control valve. The influence of the disturbance (e.g., increase in BOD or liquid flow) is sensed by the controller through the change in DO. In a feedback control system, a deviation or error in the controlled variable, DO in this case, must occur before any control action can be taken. The control system determines how much deviation will occur before the corrective action compensates for the disturbance. Figure 6.3 shows both a feedback and feedforward controller. The feedforward element senses the disturbance directly, and takes corrective action immediately. The feedback element trims the corrective action at a later time, because of inaccuracies in the feedforward element.

Figures 6.4, 5, and 6 are of the same control system, but specifically for an activated sludge plant. Figure 6.4 shows the uncontrolled case. Figure 6.5 shows a feedback controller. Information flow (i.e., DO concentration from a sensor) and physical parameters (e.g., blower

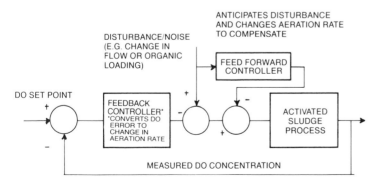

FIGURE 6.4—Uncontrolled process; Activated sludge.

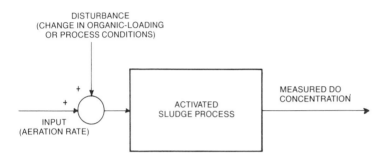

FIGURE 6.5—Feedback controlled process; Activated sludge.

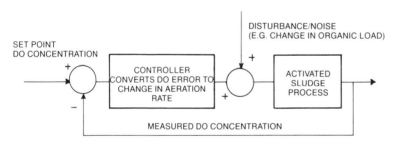

FIGURE 6.6—Feedback/feedforward controller; Activated sludge.

speed) are treated similarly on the control diagram. Each block has the ability to change units or signal magnitude. Figure 6.6 shows a feedforward controller. The feedforward controller is a device that senses or anticipates changes in organic-loading rate, then supplies a corrective signal to the aeration rate to alleviate the anticipated effect of the organic load change. The feedback controller trims the corrective signal at a later time.

Sensors are not specifically shown on the control diagrams in Figures 6.4, 5, and 6, but they must exist. For example, the feedback line in Figures 6.5 and 6.6 must contain a sensor to detect DO in the aeration process. Sensors are often included in control diagrams when their performance affects the ability to control.

The control systems shown in Figures 6.1–6.6 are simple, single-loop, control systems. In practice, multiple loops are often required. For example, many municipal treatment plants use plug-flow aeration basins. These basins usually require tapered aeration, which complicates automatic control. A control loop must exist for the inlet part of the basin, the exit part of the basin, and perhaps several sections in between, depending on the control precision needed. Figure 6.7 shows a plug-flow reactor that needs three separate control loops because of spatial gradients in oxygen demand, DO, and aeration rate. For plug-flow tanks, more than one DO sensor and control element will often be necessary. Furthermore, problems arise when using common elements, such as blowers and air piping. For example, a flow increase to grid 1 may reduce the flow to other grids. The reduced flow to other grids will cause a decrease in DO concentration; the other control systems will attempt to compensate, which may reduce air flow to grid 1. In the worst case, the tank will always be out of balance, with all three controllers "hunting." Constant valve adjustment quickly wears them out and frustrates the operators. The operators may then override the control system, concluding that either it does not work properly or that DO control was never necessary.

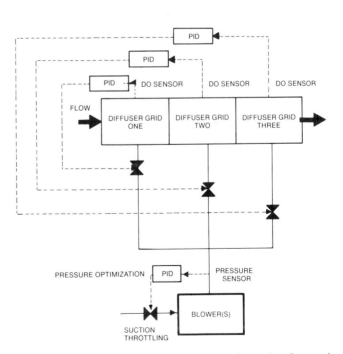

FIGURE 6.7—Three-stage DO control system for a plug-flow application.

The solution is to provide a second control loop to maintain constant blower discharge pressure, preferably by adjusting guide vanes or section valves. This ensures that sufficient air flow will exist for all three grids. The second control loop is also necessary to maintain low-pressure drop in air lines, relative to valve pressure drops.

A control system performs only as well as the weakest link. The control system cannot perform if a sensor is malfunctioning, or if it cannot manipulate aeration rate. Sensors and transducers are frequently the cause of control system failure.

STABILITY

Control systems can have stability problems. Stability problems occur when the control system performs the opposite of the needed control action. Instability in control systems has a precise mathematical definition, but in practical control systems, it means that the controller loses its set point and saturates at the high or low end of its range. Oscillation can also occur, which could cause the control system to continually hunt for the set point by increasing aeration rate, passing the set point (i.e., overshooting), then decreasing the aeration rate, and undershooting the set point. Either performance is unacceptable in a control system. Control system instability can cause rapid equipment failure or excessive wear, and also it can lead to the process failure.

A plant with multiple aeration basins will serve as an example of a stability problem. A frequent problem when changing air flow to one section of a process, or when changing the total air flow, is production of an imbalanced air flow. For example, in many large activated sludge plants, there are parallel aeration basins. Small differences in influent wastewater or recycled sludge-flow rates can produce imbalanced DO concentrations; the control system will then attempt to correct the imbalance. Either the blower system pressure then may change or the pressure in the distribution piping may change, which will modify the air distribution. If the modified air distribution is not corrected, the DO concentration over a period of 10 to 30 minutes will change, requiring yet another change by the controller. Depending on the control system characteristics, subsequent changes may grow smaller or may be sufficiently large to cause excessive equipment wear. In the worst case, which unfortunately occurs most frequently, the control system continuously changes flows and never reaches a stable point. In such cases, the control system is often abandoned and either manual control or no control is practiced. The control system becomes an unsuccessful experiment.

Establishing a way for the control system to sense the flows at each point in an air distribution system before making a change is one way to avoid this hunting phenomena. The change is then made to specified flows; changes in other flows are sensed and corrected. In this way, the control system can respond quickly to trim the undesired perturbation in flow, instead of waiting to see the effects of such perturbations in the DO concentration.

The difficulty in designing such a control system involves the sensors. Accurate flow sensors can be installed for the design air-flow rate, but they are often unstable or problematic at reduced or increased air flows.

CONTROL LAWS

A control law is an equation or procedure that translates the error signal into a control action. There are many control laws that can be incorporated into a controller. The simplest is an on and off control. In the case of a DO controller, the aeration device is turned off above the set point and turned on below the set point. Such control systems perform well when the process controlled reacts slowly, and when the mechanical equipment can be turned off and on without damage. An extension of this concept is a differential gap or dead band controller. In this case, the turn-off point is above the set point and the turn-on point is below the set point. Between the turn-on and turn-off points, the controller takes no action. A hot-water heater is an example of a dead band controller.

Because aeration system response is too rapid for on and off controllers, some type of continuous controller is required. Proportional, integral, and derivative controllers are often used. The signal manipulating the aeration rate is proportional to the error, the time integral of the error, and the time derivative of the error. The control law can be stated as follows:

$$\text{Control action signal} = K \left(E + \frac{1}{T_R} E \, dt + T_D \frac{dE}{dt} \right) \qquad (1)$$

Where;

E = error signal (set point minus measured value), and,
K, T_R, T_D = controller gain, reset time, and derivative time.

The gain and controller constants can be set empirically or can be determined theoretically for a given process through a mathematical model. Generally, there exists an optimum set of parameters, but in practice the optimum is seldom obtained. Obtaining the optimum parameters is difficult because of changing plant conditions or operation strategies.

Often controllers contain only one or two of the actions shown in Equation 1. Controllers not requiring quick response do not need derivative action. Integral action (often called reset) can compensate for offset (i.e., steady-state error), which occurs when only proportional control action is present.

Self-contained controllers are available to perform the control functions shown in Equation 1. Continuous control can be implemented through such devices without explicitly understanding Equation 1. There exists a wide variety of pneumatic, analog, and digital controllers.

*I*NSTRUMENTATION

As previously indicated, sensors or instruments are needed to implement control laws. In the case of DO control, a DO sensor of some sort is the primary sensor. A cause, if not the most frequent cause, of DO control system failure is the poor performance of DO monitoring instrumentation. This poor performance can most often be attributed to faulty equipment, improper application, and lack of attention and maintenance by plant personnel.

DISSOLVED OXYGEN SENSORS

Kulin *et al.*[6] have described a DO field protocol that can be used to select, maintain, and locate DO sensors. They evaluated equipment from several different manufacturers over a 60-day period, and they made specific recommendations with respect to probe location, calibration, and maintenance.

Probe calibration. A daily check by comparing process probes to a portable calibrated reference probe generally is recommended. The procedure typically would include calibrating the reference probe by saturating clarified process effluent, measuring this concentration with the Winkler procedure, and setting the probe to read this value. Through this method, the Beta factor is automatically included in probe calibration. For rapid calibration, air calibration in water-saturated air frequently is recommended. As a further alternative, saturating distilled or tap water, measuring DO concentration with the Winkler method, and then setting probe calibration accordingly also can be used. Any of these methods can be used and can provide calibration accuracy to within a few percent, which is adequate for automatic control purposes. One potential pitfall when using industrial effluents is interference with the Winkler procedure. Industrial effluents, even highly treated, can still contain significant interferences that cause erroneously low Winkler results.

Probe mounting. Specific mounting instructions are recommended. Support structures should be smooth without angles, turns, or other geometric deviations that could collect material. They also recommend that probes be exposed directly to the mixed liquor as opposed to mounting them in flow-through cells. Mounting hardware should be of the quick-release type that provides the greatest facility for probe maintenance and membrane replacement. Probes should be mounted so that flow currents are not perpendicular to the membrane surface; those mounted in downward flowing currents are desirable.

Probe maintenance. Kulin *et al.* do not generalize on probe maintenance requirements, but they comment that each application and manufacturer probe will have different maintenance requirements.

They place emphasis on installations that avoid fouling and abrasion on the membrane, and that facilitate maintenance. A liberal probe maintenance and calibration budget (perhaps daily) should be provided. Probe location is important not only for maintenance considerations, but also for process-specific conditions. Plug-flow tanks require several probes to properly determine DO concentration. Complete mix basins, especially those aerated by mechanical aerators, have significant DO gradients that can influence probe location.

FLOW-MEASURING DEVICES.

Flow-measuring devices for liquid and gas streams are necessary for control. Gas-flow measuring devices are particularly important. It is necessary to balance air flow to several places in an aeration basin. This requires precise, real-time flow measurements.

"Averaging pitot tubes" and orifice plates are reliable and accurate ways of measuring gas-flow rates. Unfortunately, they have a limited dynamic range. At low-flow rates, the pressure differential is too small to measure. At high-flow rates, particularly with orifice plates, the pressure drop may become excessive, making accuracy (and power saving) problematic. Consequently, the instrumentation must be designed to work over a limited range. This range must be selected for anticipated flow rates and process loadings. Because treatment plants are often designed for future flow rates that are not realized until well into the plant's life, the instrumentation and valves are sized incorrectly for a large portion of the plant's life. In many cases, providing different instrumentation for different phases of the plant's life will be necessary.

The liquid flow-measuring devices used in many feedforward applications are more problematic than gas-flow measuring devices because liquid flows usually have solids and other materials that can obstruct them. Weirs are effective for flow measurement, but they often produce erroneous results because of level adjustments. Devices that measure the level behind a weir are usually exposed to harsh and corrosive environments. Lack of maintenance frequently causes level sensors to become inoperative. At this writing, there are numerous new liquid level-measuring devices (e.g., ultrasonic detectors) that are being readily applied to wastewater applications.

OTHER SENSORS

Pressure-measuring devices are also needed in DO control systems, but they are less important than are DO sensors and flow sensors. One important exception is for measuring blower discharge pressure. Precise and accurate blower discharge pressure is necessary to ensure that operation conditions producing surge or motor overload conditions are avoided. Alarms to avoid these conditions, and automatic shut-down systems in the case of surge or overload, are essential. Accurate blower discharge pressure is also required to minimize energy consumption.

Excellent references are available on instrumentation in general[8] and for specific devices.[7,8,11,12,13] The most important concept in instrumentation selection is providing instruments suitable for the harsh treatment plant environment, while simultaneously providing a wide dynamic operating range and repeatable measurements.

METHODS OF ADJUSTING AERATION RATE

Aeration equipment that adjusts aeration rate and that conserves energy is an essential ingredient in a DO control system. Methods previously used include suction throttling of blowers, guide vane adjustments, variable- or multiple-speed motors, changing the number of units in service, and adjusting impeller submergence. Each method has particular advantages and disadvantages.

Blower throttling is often a successful method of regulating aeration rate. Changing the number of blowers in service is effective, but often this cannot be automated. Large blower start-up requires care, and most plant operators will not trust an automatic control system to start large and expensive equipment. Blowers should be selected to provide a range of operating conditions. Selecting at least two blower sizes to provide greater turn-up and turn-down capability is desirable. Blower size is more important at small plants that usually have fewer blowers and greater load variation.

Throttling or adjusting guide vanes is easier to automate. The chapter on blowers discusses two ways of reducing centrifugal blower discharge rate up to 50% by adjusting inlet guide vanes, exit diffuser, or rotational speed. Parallel operation of multiple blowers is also discussed. Variable- or multiple-speed motors can be used with positive displacement blowers. Two-speed motors are relatively inexpensive, requiring inexpensive controls. An economical configuration can provide 1 to $\frac{1}{3}$ or 1 to $\frac{1}{2}$ turndown capacity.

Changing the number of mechanical aerators in service is useful in many situations. As with parallel blowers, this technique is difficult to automate; however, starting and stopping aerators, especially high-speed aerators, can be automated. Care must be taken to avoid frivolous starting and stopping or control system oscillation. Mixing must be sufficient; in some instances, mixing energy requirements can exceed aeration energy requirements.

Variable speed motors are also available. Because variable frequency drives are available, continuously variable speed is possible. Multiple windings can provide two or more speeds. These drives are efficient over a wide range, but they are often very expensive. They can be automated and can be more useful than turning off aerators completely, because mixing can still be provided. Stukenberg[9] found that incidences of gear reducers' failures in mechanical aerators are not greater with two-speed motors than with single-speed motors.

Aeration rate can be varied with pier-mounted, low-speed mechanical aerators equipped with liquid level sensitive impellers. Liquid level is adjusted by a hydraulically adjusted weir. Stukenberg and Wahbeh[10] concluded that adjusting liquid level was a valid method of adjusting low-speed, mechanical aerator power consumption.

The design engineer faces a dilemma when specifying the method of changing aeration rate. Simple methods such as changing the number of units in service cannot provide the energy savings of more sophisticated techniques. Sophisticated equipment usually requires more frequent maintenance by highly trained personnel. Large treatment agencies usually can justify highly trained and specialized personnel; however, small plants, often in rural or isolated areas, cannot. Therefore, the design engineer must anticipate the competency level of the operators and their willingness to provide maintenance.

Turn-down or turn-up requirements must be established on a site-by-site basis. Plants that serve small cities generally have greater diurnal variation, but they have less economic incentive to provide for DO control. Variations in mass BOD_5 loading of 7 to 1 have been observed. Large installations generally have less variation in BOD_5 mass loading. From reviewing the methods available for adjusting energy input, a turndown ratio of at least 2-to-1 seems desirable and possible.

CASE HISTORIES

Genthe, Roesler, and Bracken[2] provide a review of 12 installations practicing DO control. Their review spans plant sizes from 0.044 m³/s (1.0 mgd) to 7.0 m³/s (160 mgd). Most of their case studies are comparisons between automatic and manual control systems. Comparisons are made over periods of time ranging from 24 hours to 4 months. Significant energy savings are reported for the successful designs. Unsuccessful designs often waste energy when compared to the uncontrolled case. The authors provide process flow and instrumentation diagrams for the facilities, as well as a brief description of each evaluation. Table 6.1 summarizes their results. Problems and design errors found at

Table 6.1.   Percent improvement of automatic control over manual control.

Plant	BOD removed/ air supplied blower kWh, kg/kWh	Air supplied/unit quantity of BOD removal, m³/kg	Air supplied/unit volume of influent m³/m	Change in BOD removal efficiency, %
Renton, Wash.	57.6	37.0	12	13
Palo Alto, Calif.	14.5	14.7	0	0.3
Rye Meads, England	—	23.9	23.5	0.6
City of Oxford, England	20	—	—	—
Valley Community, Calif.	− 29.7	− 32.1	− 16.4	1.1
Reno-Sparks, Nev.	− 24.2	− 31.2	9.2	− 14.6
Reno-Sparks, Nev.	58.3	40.7	18.2	14.7
Simi Valley, Calif.	10	9.7	2.6	− 1.5
Long Beach, Calif.	—	3.5	6.9	− 0.1
San Jose-Santa Clara Calif.	—	12.4	10.3	0.5
Cranston, R.I.	—	26	19.7	3.7
Cranston, R.I.	—	29	11.2	0.9

the 12 plants are summarized and their results are briefly reported in the following text.

LACK OF INTEREST OR NEED. Many of the plants do not regularly practice automatic DO control because plant-operating personnel indicated no need or interest. If the plant is meeting standards or if the standards are not being enforced, there is little interest in providing improved control.

BLOWER THROTTLING. Some plants are unable to throttle blowers because of interference with other plant processes or because the design engineer fails to provide some critical control system component.

COMPLEXITY. Plant personnel become frustrated with complex control systems and abandon them.

OVERSIZED CONTROL VALVES. Some plants have butterfly control valves that cannot effectively throttle blower discharge rate. They are either too large (e.g., closing the valve to 90% produced no change in flow, but closing the remaining 10% turned off the air flow entirely) or they cannot be positioned accurately. This results in inaccurate and sensitive positioning, which makes the control system hunt and wander.

INADEQUATE MIXING. Some designs do not provide adequate mixing at reduced aeration rates. Consequently, the aeration system operates at a high level, regardless of DO, just to maintain mixing.

Genthe, Roesler, and Bracken[2] concluded that, for 9 of the 12 plants, significant savings in air supplied per unit of flow or BOD_5 is obtained. Eleven plants show improved BOD_5 removal efficiency and five report energy savings. More consideration should be given to blower selection, process air stream interdependence, flow and loading variability, in-plant maintenance, personnel expertise, and flexibility. Not all plants need automatic DO control.

CONCLUSIONS. For wastewater treatment plants with varying loading rates, automatic DO control systems will decrease effluent variability, improve mean treatment efficiency, and conserve energy. Most existing plant designs can be improved by providing emphasis on control system dynamics, personnel, and equipment.

REFERENCES

1. Fertik, H., Paper presented at the Annual Meeting, Water Pollut. Control Fed., Atlanta, Ga. (1983).
2. Genthe, W.K., Roesler, J.F., and Bracken, B.D., "Case Histories of Automatic Control of Dissolved Oxygen." *J. Water Pollut. Control Fed.*, **51,** 2257 (1978).
3. Stenstrom, M.K., and Poduska, R.A., "The Effect of Dissolved Oxygen Concentration on Nitrification." *Water Res.* (G.B.) 643 (1980).
4. Roesler, J.F., "Plant Performance Using Automatic Dissolved Oxygen Control." *J. Environ. Eng. Div., Proc. Am. Soc. Civ. Eng.*, **100,** 1069 (1974).
5. Jenkins, D., Richard, M.G., and Neethling, J.B., "Causes and Control of Activated Sludge Bulking." Paper presented at the Institute of Water Pollution Control, Port Elizabeth, Republic of South Africa (1983).

6. Ogata, K., "Modern Control Engineering." Prentice-Hall, Englewood Cliffs, N.J. (1970).

7. Kulin, G., Schuk, W.W., and Kugelman, I.J., "Evaluation of a Dissolved Oxygen Field Test Protocol." *J. Water Pollut. Control Fed.*, **55,** 178 (1983).

8. Norton, H.N., "Handbook of Transducers for Electronic Measuring Systems." Prentice-Hall, Englewood Cliffs, N.J. (1969).

9. Stukenberg, J.R., "Physical Aspects of Surface Aerator Design." *J. Water Pollut. Control Fed.*, **56,** 1014 (1984).

10. Stukenberg, J.R. and Wahbeh, V.N., "Surface Aeration Equipment: Field Testing versus Shop Testing." *J. Water Pollut. Control Fed.*, **50,** 2677 (1978).

11. "Instrumentation in Wastewater Treatment Plants." Manual of Practice No. 21, Water Pollut. Control Fed., Alexandria, Va. (1987).

12. "Process Instrumentation and Control Systems." Manual of Practice No. OM-6, Water Pollut. Control Fed., Alexandria, Va. (1984).

13. "Energy Conservation in the Design and Operation of Wastewater Treatment Facilities." Manual of Practice No. FD-2, Water Pollut. Control Fed., Alexandria, Va. (1982).

Chapter 7

Operation and Maintenance

*I*NTRODUCTION

This manual was written from a design viewpoint and the O & M considerations affecting design have been addressed in the system-specific chapters. This chapter provides additional information to assist the design engineer in related efforts such as preparing O & M manuals, providing assistance during startup, and troubleshooting.

The following sections discuss system monitoring and recordkeeping, startup, normal operation, shutdown, preventive maintenance and troubleshooting. All of the sections are written specifically for applications that involve the activated sludge process. Most aeration system applications are for this process; the cost of energy required to run the process is always a significant portion of the plant's operating budget. Although the information presented will apply to other applications, the level of attention required for the other applications (other than for maintenance) will generally be less extensive.

The sections on system monitoring and recordkeeping cover the general requirements for any aeration system and the specific requirements for diffused air and mechanical-aerator systems. Separate sections are provided for each discussion for diffused-air systems and mechanical-aerator systems.

*S*YSTEM MONITORING

Within the constraints placed on the activated sludge system, the major objective will be to transfer the needed mass of oxygen using as little energy as possible to operate the system (i.e., maximize the aeration efficiency that is defined as the mass of oxygen transferred per unit of total power input). Accomplishing this goal requires knowing how various operating parameters affect the OTR and the total input power.

The OTR of wastewater can be affected by changes in process parameters such as solids retention time and DO concentration because of changes in diffuser performance characteristics with time of use in addition to the effects of fouling.

For diffused air systems, all the power is drawn by the blowers used to supply the air to the diffuser system. Therefore, any factor that increases either the volume of air to be compressed or the discharge pressure will adversely affect the aeration efficiency. More air will be required if the OTE (mass of oxygen transferred per unit mass of oxygen supplied) decreases. Operation at the higher air flow increases the diffuser flux, which increases mean bubble size, with consequent further loss of efficiency. Supplying more air to make up for lost efficiency also results in a higher discharge pressure because line losses will increase. The aeration power required can also increase without any decrease in OTE if the pressure drop across the diffusers increases because of clogging. Air-side fouling and liquid-side inorganic fouling of fine pore diffusers can actually increase the OTE; however, attendant increases in pressure drop usually result in reduced aeration efficiency. Liquid-side fouling by biological slimes usually results in lower OTE and sometimes higher pressure drop across the diffuser element.

For mechanical aerators, the aeration efficiency is principally affected by changes in the OTR. (Rotational speed and direction may sometimes be a problem.) The key equipment-operating parameters for mechanical aerators are those that affect the aerator impeller such as submergence, rotation rate, ragging, other types of fouling and wear.

Rigorous methods for measuring OTE and OTR under process conditions are available (see Chapter 8). One or more methods may be appropriate for a particular aeration system; however, the rigorous methods typically are time consuming and therefore may be too costly for use in day-to-day monitoring. As an alternative, calculated ratios of operating data can provide good indications of overall system efficiency. If the monitoring program indicates a need for more detailed data, a more sophisticated rigorous test can be performed.

The above monitoring program, which is not considered applicable for absolute evaluations of different systems, is considered useful in evaluating relative performance within a given system which is consistently operated. The suggested ratio for monitoring the overall aeration efficiency of diffused air systems is the volume of air supplied per unit mass of BOD removed. This measurement should be made with care as air density is sensitive to temperature and pressure. If nitrification is taking place, the mass of ammonia nitrogen oxidized to nitrate must also be monitored. Units of measurement must be consistent for both nitrification and BOD. A second ratio applicable to both diffused air and mechanical aeration systems is the electrical power consumed per unit mass of BOD removed. Because the transfer efficiency is also affected by the DO concentration, the DO concentration in the mixed liquor must also be monitored. Multiple-day, running averages can be used to reduce daily fluctuations in the calculated parameters.

An effective and quick way to evaluate changes in the aeration efficiency is to plot air rate (or power usage) per kilogram of BOD removed (plus oxygen consumed in nitrification if applicable) and DO versus time. Such plots are commonly known as trend charts.

If an increasing trend of air rate/kg organic removal is observed, an increase in the oxygen demand per unit mass of BOD removed or a decrease in the aeration efficiency is indicated. Factors affecting oxygen demand include the characteristics of the incoming wastewater and operating parameters such as solids retention time. These factors and equipment-operating parameters also affect OTE. If an increasing trend is observed, additional evaluations are required. More discussion on the effects of equipment operating parameters are presented in the Diffused Air and Mechanical Aerator troubleshooting sections of this chapter.

In addition to the measured parameters, visual observations should be made at least daily. For diffused-air systems with the diffusers laid out in a grid pattern, the surface turbulence should be uniform. The surface pattern should be free of boiling. Boiling usually indicates leaks in the submerged air supply piping (i.e., leaking gaskets, faulty joints, and broken diffusers). For fine pore diffusers, the size of the bubbles exiting the surface of the liquid should be noted. If the bubbles increase in size as the time in service increases, fouling of the diffusers may be occurring and OTE declining. In plug-flow aeration tanks, the fouling may not be uniform throughout the tank; therefore, observations should be made at several locations.

For mechanical-aerator systems, the surface patterns should also be observed. Surging in basins equipped with multiple-surface aerators can not only reduce the oxygen transfer capability, it can also cause damage to aerator shafts, impellers, and reducer components (e.g., bearings). For sparged-turbine aeration systems, the surface pattern should be relatively uniform. There should not be any geysers caused by leaks or breaks in the air supply piping. The operators should walk through the plant daily and should listen for unusual noises from rotating equipment such as blowers and gear boxes on mechanical aerators and for air leaks in the air-distribution system.

Monitoring of the aeration system should be done regularly. More frequent monitoring during initial operation will provide baseline data from which changes in system performance can be evaluated.

RECORDKEEPING

The following operations data should be recorded on a regular basis. Daily collection and recording is recommended at least initially. A change in the frequency to more or less often can be made once operating experience is gained.

1. Weather conditions. Temperature, barometric pressure, humidity, wind direction, and precipitation;
2. Aeration tanks. Visual observations of the mixed liquor especially the surface pattern. For systems using compressed air, look for geysers and listen for air leaks in the distribution piping. Observe the bubble size for fine-pore, diffuser systems. For mechanical aerators, listen for unusual noises and watch for unusual vibrations. The spray patterns should be noted as an indication of change in submergence; and
3. Process operation. Note any changes in operation such as shutdowns, solids wasting rates, and unusual wastewater characteristics.
4. Other recordings of significance include SRT, DO, BOD loading and power consumption.

For diffused-air systems, the following data should also be recorded.

1. Blower operation. Discharge air-flow rate, pressure and temperature, time in service, oil pressures, vibration, bearing temperatures, and power consumption. The power drawn should be measured using a multiphase wattmeter. When using multiple blowers, power factor should be measured too;
2. Air-filter conditions. Time in service and differential pressure;
3. Air-distribution system. Total air-flow rate, air-flow rate to each tank, and line pressures at the aeration tanks and other key points in the system. When travelling about the plant, listen for air leaks and see that they get fixed. A significant amount of energy can be wasted by allowing air to leak to the atmosphere rather than being delivered to the process;
4. Condensate blowoffs. Date operated, estimate of volume removed, and quality (clarity); and,

For mechanical aerators, the following data should be recorded.

1. Aerator conditions. Time in service, power consumption (if multiple units are used the total power drawn should be measured and

recorded), any unusual noises or vibrations, and speed (if adjustable); and,

2. For combination-diffused air and mechanical-aerator systems, the records suggested for both types of aeration systems should be kept.

Other process records are required. For activated sludge systems, the data would include wastewater, return sludge, and waste sludge flow rates, and concentrations of influent and effluent BOD, mixed liquor, return sludge and waste sludge suspended solids, mixed liquor DO concentration and temperature. Note special events such as power outages and their duration.

In addition to the operations data, all maintenance work should be documented. Suggested maintenance records are detailed in Manual of Practice, OM-3, Plant Maintenance Program.[1]

The results of all special studies such as wastewater characterizations, diffuser-cleaning tests, and oxygen transfer tests should always be written up in a concise report and filed for future reference. The report should include the purpose of the study, the methods used, the results, and the conclusions. All raw data and calculations should be appended to the report.

DIFFUSED AIR SYSTEMS

This section discusses the startup, normal operation, shutdown, preventive maintenance, and troubleshooting of diffused-air systems.

STARTUP

Before starting up any rotating equipment such as the blowers, the equipment should be properly checked out and lubricated according to the manufacturer's instructions. If a basin is put into service during cold weather and an ice layer has formed, buoyant forces exerted by ice trying to float when the basin is filled could cause damage.

The following steps should be followed whenever putting into service an aeration basin equipped with a diffused-aeration system:

1. Check the air piping and diffuser system for loose joints, cracked piping and other defects. Leaks or faulty joints that will not stand up under normal operating stresses could result in wastewater solids entering the system after startup and lead to air-side fouling. If the system is equipped with pressure taps and tubing for monitoring dynamic-wet pressure, make sure all tubing connections are tight.

2. Start feeding air to the diffuser system before the diffusers become submerged. For fine-pore diffusers, always feed at least the manufacturer's minimum recommended flow rate per diffuser. This will prevent backflow of wastewater into the diffusers. Although less critical, coarse-bubble diffuser systems should also be fed enough air to keep wastewater out of the system.

3. Fill the aeration basin to a level about 30 cm above the diffusers. Observe the air distribution and check to make sure there are no significant leaks. If service water is available, it is preferable to wastewater or mixed liquor for the initial filling. Have any leaks repaired before continuing to fill the basin. Use caution during the early stages of filling so the force of incoming liquids does not damage the diffuser system or its supports.

4. Continue to fill the aeration basin while monitoring and adjusting the air-feed rate. Unless it is adjusted, the air feed rate will

decrease and discharge pressure will increase as the liquid level in the basin rises.

5. On fine-pore diffuser systems, operate the condensate blowoffs one at a time until the system is free of moisture.

6. Adjust the flow rate of mixed liquor, wastewater, and air to the basin to meet the desired process-operating conditions.

NORMAL OPERATION

The aeration system will require monitoring to maintain acceptable performance of the process and the air-diffusion system. The surface pattern in the aeration basins should be visually observed at least daily. The surface patterns generated by different types of aeration systems and configurations will vary. The operator should become familiar with the system so unusual circumstances can be picked out. For example, localized turbulence may indicate damaged piping or diffusers.

Generally, it is desirable to keep a relatively uniform DO concentration throughout the aeration basin. DO concentrations will change with depth, however. If the basin operates in the plug-flow mode and the BOD loading to the basin is relatively high, more air will be required at the influent end than at the effluent end. When adjusting the air-flow rates, the diffusers must operate within their proper range of air-flow rates. Too low an air rate can result in poor distribution and backflow of wastewater into the diffuser holder. Too high an air rate can cause excessively high pressure drops across the diffusers and coarse bubbling.

If the air rate required for process needs exceeds the diffusion-system capacity, adjustments will be required. More aeration basins should be put into service if the air rates to all portions of the basin are high. If the air rate per diffuser problem is localized, diffusers may have to be added to or removed from the system. Diffusers are added to provide more transfer capacity where the loading is high and diffusers are removed to avoid overaeration where the loading is low. Extra diffuser holders or taps can be provided as part of the design. This makes adding diffusers easy after start-up. When removing diffusers, the air rate to that portion of the basin must not be reduced below the acceptable level required for mixing.

On fine-pore diffusion systems, the condensate blowoffs should be operated on a set schedule. Because the amount of condensate depends on the season and weather conditions, plant-operating experience will determine the frequency needed. Initially, weekly checks should be made. When blowing off the condensate, sample and inspect the liquid removed. The liquid should be clear. If it contains solids, there may be a leak in the submerged portions of the air piping. The piping should be inspected and repairs should be made as soon as practical.

On fine-pore aeration systems provided with dynamic wet-pressure–monitoring equipment, measurements should be made and recorded weekly for the first 6 months of operation to provide baseline data and to determine if fouling occurs rapidly. Data generally are collected at the same air-flow rate per diffuser. After the first 6 months, the frequency of data collection can be reduced to monthly.

SHUTDOWN

If an aeration basin must stand idle for more than 2 weeks, it should be drained and thoroughly cleaned. Once cleaned, the aeration basin should be filled at least partially with clean water. Although these systems should be designed to withstand the range of temperatures encountered in the region, observance of the above precaution will result in the maintenance of a favorable safety factor from the standpoint of thermal expansion and contraction. Groundwater levels and basin buoyancy must be considered. Aeration basins should not be drained during freezing weather unless absolutely necessary. Ice build-up and frost heave can cause serious damage. Ice build-up can be alleviated by bubbling air into the basin thus preventing serious ice damage. Covering the basin floor with straw or providing heat may be necessary to prevent frost heave.

If a basin is removed from service during freezing weather, it can be idled by feeding the greater of either the minimum recommended air rate per diffuser or the minimum air rate required to keep the solids in suspension. When taking an aeration basin out of service for more than 2 weeks, these steps should be followed:

1. Stop the wastewater and return sludge flow to the basin but continue to feed air to the diffusers at a minimum rate.
2. Open drain lines and start the drain pumps if necessary. Continue to feed air to the system until the liquid level is below the washed-off diffusers. Monitor and adjust the air rate as the liquid level falls.
3. Once the basin is drained, check to make sure that the ground-water pressure relief valves are operational so uplift pressures do not cause damage.
4. Washdown the basin walls and floor, the air piping, and the diffusers to avoid odor problems. Materials accumulated on fine-pore diffusers either should be removed, or at least not be allowed to dry.
5. Inspect the air piping and diffusers. Check to ascertain that gaskets requiring externally applied sealing force are adequately adjusted. Make repairs as needed.
6. For above freezing conditions, fill the basin with clean water to an elevation about 1 m (3 ft) above the diffusers. Fine pore-diffusers must be fed air at a rate equal to or greater than the manufacturer's recommended minimum. For freezing conditions more water may be needed to protect other normally submerged piping. Although the air being fed will normally prevent serious ice damage, if an ice layer does form, do not drain the water from the basin. If the ice layer were to break and fall on the air-diffusion system, serious damage could result.

If the basin is to be taken out of service to effect repairs that can be completed in a few hours, no special cleaning is required except in the work area. If the work requires several days to complete, the basin and diffusion system should be washed down.

PREVENTIVE MAINTENANCE

The preventive maintenance system should include the major aeration equipment—blowers, air distribution and filtering system, and diffusers.

Blowers. The manufacturer's recommended maintenance requirements should be followed to minimize blower problems. Positive-displacement blowers require periodic lubrication. The small units have grease-type ball bearings; the larger units usually have a positive-feed oil system with a pressure pump. Oil should be changed as recommended. At intervals of 2 to 4 years, the blower end plates should be inspected for wear and replaced or restored to original clearance. Oil from the blowers can be analyzed to determine reasons for oil fouling or potential wear problems.

Turbine or centrifugal blowers when used with an electric motor require a minimum of attention. The manufacturer's maintenance manuals should be consulted for lubrication schedules, rotor balancing, and noise control.

Regardless of the blower type in use, daily inspections should be made to check the blower operation for excessive vibration, overheating, and unusual noises.

Air System. The air system contains the filtration equipment, the air-distribution piping and the air-flow instrumentation. The filtration-equipment maintenance consists of cleaning and changing filter media, and cleaning of ionizer elements in electrostatic-filtration units. The manufacturer's recommendations for maximum head loss should be used to gauge when filter elements must be cleaned or changed. The ionizer elements of an electrostatic-filtration unit must be wiped down about every 2000 hours.

Air-distribution piping generally requires very little maintenance. Inspection and repair of protective coatings and joint gaskets is usually all that is required. The entire system should be checked for leaks at least annually. The aeration efficiency can significantly deteriorate if there are leaks that rob air from the process.

Because accurate air-flow measurement is essential for monitoring a diffused-aeration system, calibration of the air-flow meters is an important maintenance task. A full calibration should be performed when the meter is first put into operation. The calibration should be checked after the meter has been in service for 1 month. If no adjustments are required, the time between calibration checks can be increased until the proper frequency is determined.

Diffusers. The appropriate preventive maintenance procedure and frequency depends on the system provided and the service conditions.

Typically, coarse-bubble diffusers require very little preventive maintenance. Some slime and built-up deposits may be dislodged from the diffusers by air bumping the system. The procedure is to increase the air rate to the system for several minutes. The increased air rate should be determined by manufacturer's recommendations and by system capabilities. If the diffusers are mounted on fixed headers, the aeration basins should be drained annually for inspection of the diffusers and submerged piping even if the system seems to be operating satisfactorily. Accumulated trash and stringy materials, weakened support brackets, and partial plugging of orifices can be remedied more easily if this preventive maintenance is done under favorable conditions of weather and wastewater flow.

Fine-pore diffusers are much more susceptible to fouling than coarse-bubble units, so preventive maintenance that is aimed at keeping the diffusers clean and efficient is very important. The diffusers have to be removed from the basins for some cleaning methods but can remain in place for other methods. The in-place methods can be further divided into two categories, those requiring process interruption (e.g., basin draining) and those that can be performed without interrupting the process operation. Methods requiring diffuser removal include refiring of ceramic elements, acid soaking, treatment with detergents, and high-pressure jetting of tube diffusers. Because the time and effort required to perform these methods is substantial, they typically are not used as preventive maintenance techniques.

Methods that require process interruption but that are performed on the in-place diffusers include hosing with either low (<400 kPa, 60 psig) or high (>400 kPa) pressure water, brushing, treatment with acid, steam cleaning and sandblasting. Many of these methods are effective and economical, especially in plants where there is reserve aeration capacity or multiple-aeration units that allow a basin to be taken out of service for cleaning. The frequency of cleaning by a process-interrupting method will be limited by economic and process considerations.

Two procedures that do not require process interruption have been recommended by manufacturers of fine-pore, diffused-air systems. They are air bumping and *in-situ* acid cleaning. The air-bumping procedure can be used on both flexible membrane and fine-pore, ceramic diffusers. Although the *in-situ* acid-cleaning system could be used on any suitably designed diffused-air system, the available systems were developed for fine-pore, ceramic and plastic diffusers.

The air-bumping procedure for flexible-membrane diffusers, also called flexing, is accomplished by shutting off the air to the diffusers and by allowing them to collapse onto the frame. The air flow is then increased to between 2 to 3 times the normal air rate for several minutes and then returned to the desired operating condition. The procedure is done on a grid-by-grid basis to minimize the air required. Flexing is typically performed every 2 to 4 weeks.

For air-bumping ceramic-disc diffusers, the air-flow rate is increased to 1.5 L/s (3 scfm) per diffuser for 15 minutes than returned to the normal operating range. This can be accomplished rather easily when the blowers are rotated. The blower being put into service is started before the blower being taken out of service is shut down. Caution must be used to avoid costly demand charges. One manufacturer recommends bumping the entire system weekly.

In-situ acid cleaning of ceramic- and plastic-diffuser systems can effectively control certain types of fouling. Two systems are available; one uses anhydrous hydrogen chloride gas and the other uses liquid formic acid. These systems are most effective on systems encountering inorganic fouling by acid-soluble compounds such as iron hydroxides, and calcium and magnesium carbonates. There is also evidence that they are effective in controlling the effects of biological fouling.

The *in-situ* acid-cleaning procedures require increasing the air-flow rate per diffuser to almost the maximum to provide good distribution and to get as many pores operating as possible. The cleaning agent is then injected into the air stream until the dynamic-wet pressure stops decreasing. With the HCl-cleaning system,[2] the gas reacts with the water entrained in the diffuser element to form liquid HCl. The acid, which reaches a concentration of about 28%, dissolves the acid-soluble deposits. Typically, a system is cleaned one grid at a time; each grid takes about 30 minutes. This procedure is covered by a United States Patent.

Because fouling of fine-pore diffusers is site specific, the system should be designed so various preventive maintenance procedures and frequencies can be evaluated after initial startup. For flexing of membrane diffusers, local air-control valves and flow meters should be provided. Ceramic- and plastic-diffuser systems can be designed to be compatible with acid cleaning. Portable acid-feed systems can be used for testing, then a permanent feed system can be provided later if the cleaning process proves to be functionally and economically attractive.

The effectiveness of the preventive maintenance program must be based on controlling increases in operating pressure and losses in OTE because both factors affect the aeration efficiency.

Changes in operating pressure should be evaluated using dynamic-wet pressure not air main pressure. For fixed grid systems, one or two diffusers per grid should be equipped with pressure taps to measure dynamic-wet pressure as shown in Figure 7.1.[3-13] A different frequency can then be used on individual grids to determine if and at what frequency the procedure is effective. Because fouling rates can vary with location in plug-flow aeration basins, the assignment of frequency to the test grids must be made carefully to avoid misleading results.

The effects on OTE can be determined relatively easily if more than one basin is in service. The preventive-maintenance technique is applied to one basin and not to the second basin. The performance of the two basins can then be evaluated by monitoring the air volume used per basin as long as the two basins are operated in parallel (i.e., equal loadings, SRT and DO concentration). A process water test of the two basins can provide the comparative performance of the two basins directly, but at somewhat greater cost. The test period should be

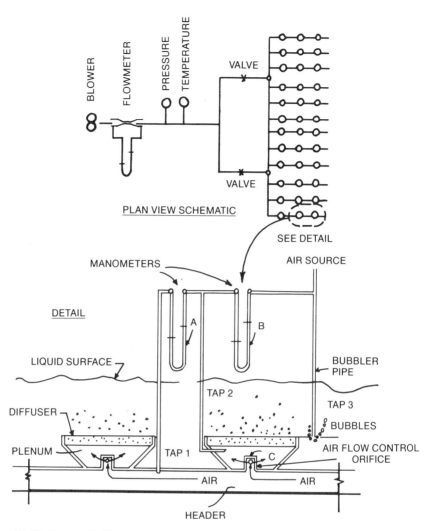

FIGURE 7.1—Diffuser pressure monitoring apparatus. Manometer "A" measures orifice pressure differential that can be correlated to air-flow rate. Manometer "B" measures dynamic-wet pressure.

long enough to separate long-term trends from short-term fluctuations. If fouling is rapid, 1 to 2 months should be adequate. If fouling rate is slow, 6 months to 1 year may be necessary. If no effects are evident after a year, the preventive maintenance procedures should be evaluated in conjunction with the annual basin draining for inspection and cleaning. The preventive-maintenance program could make cleaning the diffusers during draindown easier. For example, *in-situ* acid cleaning a fine-pore system could control fouling so chemical treatment during draindown was not needed. Methods for performing the evaluation are described in the following section.

Because wastewaters can change with time, the preventive maintenance evaluation should be repeated as specific plant experiences indicate.

TROUBLESHOOTING

Troubleshooting problems with coarse-bubble diffusers is a relatively simple task. Significant leaks in the system will usually be seen as localized turbulence on the surface of the aeration basin. Minor leaks may not be noticeable, but they will have little effect on the aeration efficiency. They can be found and corrected during the annual draining. Although fouling is generally not a problem, coarse-bubble diffusers will require occasional cleaning to remove debris and stringy materials that could affect operating pressure and OTE. Again, the annual inspection should be adequate.

Problems arising with fine-pore, diffuser systems are usually associated with fouling. The following steps are recommended for troubleshooting fouling problems:

1. Obtain a representative sample of new and used diffusers.
2. Characterize the diffusers to determine what effect the foulant has on dynamic-wet pressure and OTE.
3. Analyze the foulant.
4. Evaluate cleaning methods for restoring dynamic-wet pressure and OTE.

Samples of diffusers are readily available if the diffusers are mounted on removable headers. Diffusers from grid systems will require draining a basin unless the system was provided with some test diffusers on removable headers. When collecting samples from plug-flow aeration basins, diffusers should be taken from several locations down the length of the tank because the fouling may not be uniform.

The fouled diffusers should be tested before any cleaning methods are evaluated to quantify the degree of fouling. Several new diffusers should also be tested to provide comparison data. The following tests are recommended:

(a) dynamic-wet pressure at minimum, mid-range, and maximum recommended air rates;
(b) bubble-release vacuum (BRV) profile (fine pore diffuser elements only); and
(c) small-scale clean-water, oxygen-transfer tests.

Dynamic-wet pressure can be conducted in an aquarium or similar small tank that uses the apparatus shown in Figure 7.1 except that the static head is measured with a scale rather than a bubble tube. Dynamic-wet pressure is then calculated by subtracting the static head from the holder pressure.

BRV measures the differential pressure required to form bubbles at a localized point on the surface of a thoroughly wetted diffuser element. It provides a means of comparing relatively effective pore diameters at any point on the surface. The apparatus used for measuring BRV is shown in Figure 7.2. The procedure for conducting a BRV test is detailed elsewhere.[3]

BRV measurements are made at 8 to 12 points on the porous medium to ascertain the degree of uniformity. The uniformity is quantified by calculating the mean (BRV), the standard deviation of the mean (s) and the coefficient of variation (s/BRV). For new porous medium-diffuser elements manufactured using modern methods and appropriate quality control, the coefficient of variation should be less than 0.05. Fouled elements, even those showing little or no increase in dynamic-wet pressure, typically show significant increases in the mean BRV and in the coefficient of variation. Air-side fouling can and should be tested by checking the BRV of the air-side of several diffusers.

If the foulant buildup is sufficient to collect a sample, it should be analyzed for total solids and total volatile solids. If the foulant has a volatile content of 60 to 80%, its origin likely is principally biological. If the volatile fraction is less than 50%, a significant inorganic component is indicated. Because inorganic constituents often include iron, calcium, or magnesium compounds, it is useful to treat the residual from the total solids analysis with acid (18% HCl) to see what fraction

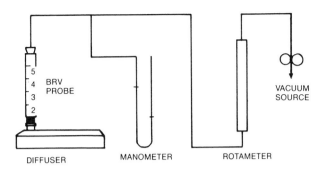

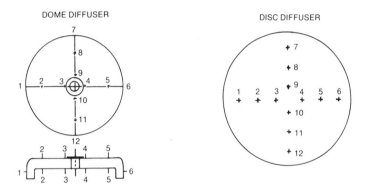

FIGURE 7.2—Bubble release vacuum BRV test apparatus and typical testing points for dome and disc diffusers.

is acid soluble. It should be noted that inorganics may also be entrained in biomass as a precipitate. This information will be useful during evaluation of cleaning methods.

Once the dirty diffusers have been characterized, the effectiveness of various cleaning methods can be evaluated. An iterative process is used whereby a cleaning method is tested and, if visual observations indicate that the method was effective, the effectiveness is quantified by once again measuring dynamic-wet pressure, BRV and OTE. The simplest cleaning method should be tested first. More involved methods can be tested later, if necessary. Biological slimes (i.e., foulants having high-volatile content) can usually be removed by mechanical means such as hosing with a low [<400 kPa (60 psig)] or high [(>400 kPa (60 psig)] pressure water spray or brushing. Inorganic precipitates firmly attached to the diffuser element may yield to chemical treatment.

Once an effective-cleaning procedure is developed in the laboratory using the small-scale tests, it can be applied to the full-scale system with confidence.

If time or process constraints will not allow such a rigorous diffuser-testing program but system monitoring indicates that cleaning is necessary, the iterative approach can be applied to the full-scale system. A portion of a grid can be used to test various cleaning methods. Cleaning using hosing to remove loosely attached materials, brushing to loosen residual materials and rehosing, and acid treatment to remove scale deposits have all been effective in several installations.[14,15,19,20] Using one or more of these steps in order is the most effective.

Between each step, the treated units must be evaluated to determine if more cleaning is needed. The evaluation can be visual observation or some test to determine quantitatively the effectiveness of the cleaning procedure. On rigid, fine-pore diffusers (e.g., ceramics or porous plastics), the BRV test can be performed on the in-place units.

If the initial hosing effectively removes all of the foulant except scale, the second step can be skipped. Before treating any diffuser element with acid, check with the manufacturer to make sure that the construction materials will not be affected adversely.

Numerous fine-pore, diffuser-cleaning methods have been tested over the years and reported in the literature. Table 7.1 summarizes several studies involving cleaning methods that required removal of the diffusers. Table 7.2 summarizes in-place cleaning studies.[4-18]

MECHANICAL AERATOR SYSTEMS

This section discusses the startup, normal operation, shutdown, preventive maintenance, and troubleshooting of mechanical and combined-mechanical aerator-diffused air systems.

STARTUPS

Before starting up any rotating equipment, the equipment must be properly checked out and lubricated according to the manufacturer's instructions. If a basin is put into service during cold weather, care must be taken to avoid damage caused by floating ice. All rotating parts must be inspected. Loose parts must be tightened and broken parts fixed as needed. Debris should be removed from the shaft and impeller. The aerator should be operated to observe for proper rotation, unusual sounds, excessive vibration, excessive heat, or oil leakage. If any of these conditions exist, the aerator should be shut down until the cause can be determined and corrected. For initial startup, the manufacturer's recommended procedures for run-in should be followed.

If possible, the basin should be filled with service water to the normal operating level and the aerators started to provide mixing before adding mixed liquor. For sparged air systems with radial-flow turbine impellers, air feed should be started before the liquid level reaches the air-sparging device to keep liquid out of the air-distribution piping.

Down-pumping, axial-flow, draft-tube aerators should have a minimum liquid level above the top of the draft tube before the mixer is started. After the proper liquid level is achieved, the mixer should be started and operated without air for a period of 30 to 45 minutes to establish stable hydraulic conditions. Air can then be introduced at or below the rate specified for the aerator. The air should be forced downward into the draft tube and should surface as fine bubbles outside the perimeter of the draft tube. If too much air is introduced, the air can overcome the downward flow of liquid and can rise directly above the draft tube, often in a geysering mode. This condition is known as impeller flooding. If flooding occurs, the aerator must be shut down immediately and restarted in the ungassed condition. When changing the aerator speed from high to low, the air rate must be adjusted first to ensure that flooding does not occur.

The aerator speed or impeller submergence should be adjusted, as should the flow rate of mixed liquor (and air if applicable), to meet the desired process-operating conditions.

NORMAL OPERATION

The aeration system will require monitoring to maintain acceptable performance of the process and of the mechanical aerators. At least daily, the surface pattern in the aeration basins should be visually

Table 7.1. Cleaning methods for fine bubble diffusers removed from the tank.

Cleaning Agent Diffuser Type	Procedure	Results and Observations
Refiring ceramic plates (aluminum-type) from Chicago Southwest plant	Same time and temperature cycles used in the original manufacturing process	The plates which were primarily air-side clogged had original permeability restored after firing. Other time and temperature cycles were not effective.[4]
Refiring ceramic plates from Cleveland Easterly plant	Plates that had been cleaned 3 times using the Wirts[7] method were refired as follows: 12 plates were treated in an electric furnace by increasing the temperature 25°C per hour up to 600°C, held 2 hours at 600°C, then cooled in 10 hours	Original permeability was not restored.[4,7]
	12 plates were treated in a regular tunnel kiln operated according to the normal 8-day cycle using a maximum temperature of 350°C	Original permeability was not restored but results were better than the abbreviated refiring cycle.
Refiring ceramic plates at the Mogdon Works, U.K.	Using the kiln installed at the plant, plates were refired by increasing the temperature to 1550°F over 4.5 hour period, holding the temperature at 1550°F for 4 hours then cooling over 24 hours	The treated plates were restored to about 85% of their original permeability. A refiring temperature of 1800°F resulted in excessive breakage.[4]
Refiring ceramic plates at Birmingham, U.K.	Details not given	Unless the furnace temperature was at least 1290°F, recovery of permeability was less than 50%. Plate losses averaged about 20%. Refiring was abandoned in favor of an acid-detergent cleaning method.[4,8]
Refiring ceramic disc diffusers from Milton, Wis.	The discs were refired at 1100°F	The dynamic wet pressure increase because of the clogging was reduced by 80%.[8]
Refiring, acid soaking, and hosing of a ceramic disc diffuser	A clogged diffuser was refired at 1350°F then soaked in concentrated HCl for 23 hours and hosed off using a 120 psig water stream	The clogged diffuser had a dynamic wet pressure of about 62 in. of water at 3.25 Scfm/ft². Refiring reduced it to about 11 in. of water. Acid soaking and hosing reduced it to about 7 in. of water.[8]
Acid soaking of ceramic plate diffusers from Cleveland Easterly plant	Diffuser plates were soaked in acid solution for several hours then scrubbed with a mixture of air and water under pressure in a specially designed scrubbing machine (Wirts method)	Clogging was partially because of iron deposits. The most effective acid was 2% sodium dichromate in concentrated sulfuric acid. Soaking and rinsing restored permeability by about 70%, scrubbing improved recovery to 91%. Cleaning at 3 to 4 year intervals was estimated.[7,9]
Acid soaking of ceramic plate and tube diffusers from Chicago Southwest plant	Diffuser plates and tubes were soaked in an acid solution for several hours, then scrubbed with a mixture of air and water drawn down through the plates in a specially designed scrubber	3000 plates and tubes clogged primarily by dirty air were cleaned primarily by the method. Permeability recovery was approximately 80%. Soaking for 48 hours in sulfuric acid gave the best results.[4,10]

Table 7.1.   (Continued)

Cleaning Agent Diffuser Type	Procedure	Results and Observations
Multiple cleaning of ceramic plates at Birmingham, U.K.	Plates were removed and inspected. Greasy plates were soaked in a sodium-metasilicate detergent, drained and soaked in a nitric-acid bath for 2 to 4 days. The plates were then rinsed, put in boiling water for 24 hours, soaked again in a detergent solution for 24 hours, and rinsed and evaluated using an approximate permeability test	The method was the result of 4 years of testing. It replaced refiring. The nitric-acid step was for removing iron deposits.[4]
Multistep cleaning of ceramic tubes from the Gary, Ind. plant	Tubes were hosed off and scrubbed in place, removed, soaked in hot 40% sulfuric acid for 20 minutes, brushed, soaked in a dry-cleaning solvent to remove grease, and hosed again with water	The tubes returned to approximately original wet pressure after each cleaning.[4]
Acid soaking of ceramic plates from the Richmond, Ind., plant	Wirts method[7]	The wet pressure recovery was complete after each cleaning.[4]
Multistep cleaning of saran wrapped tubes at the Columbus, Ohio, plant	Saran wrapped tubes were soaked in 3% sodium silicate plus a non-ionic detergent, hosed with high-pressure water, immersed for 30 second in a 10% HCl solution, and hosed off with water	The sodium silicate and detergent treatment effectively removed a silica coating while the acid treatment removed iron deposits.[10]
Caustic cleaning of ceramic plates at the Indianapolis, Ind., plant	Plates were boiled for 5 hours in 6% NaOH solution, washed thoroughly and soaked for 18 hours in 10% muriatic-acid solution	Clogging was caused by dirty air and organic growths. Cleaning was performed every 2 years.[11,12]
Caustic cleaning of plates at Mansfield, Ohio	Plates were boiled in a strong caustic solution and rinsed	Recovery in permeability was about 96%. Plates were clogged as a result of organic deposits.[11,12]
Caustic cleaning of ceramic plates at Hackensack, N.J.	Plates were soaked in a 20% NaOH solution, rinsed, drained, dipped in a concentrated nitric acid solution and rinsed	Plates were clogged as a result of organic deposits.[11,12]
Acid soaking of ceramic-tube diffusers	Soak tubes in a 1:1 solution of muriatic acid in water for at least 1 hour or as long as needed to remove deposits, then rinse thoroughly	Method is used for dissolving iron deposits.[11]
	Soak tubes in a dichromate solution made by adding 0.9 kg (2 lb) of sodium or potassium dichromate in 4.5 kg (10 lb) of water. To the dissolved dichromate, **slowly** add 45.4 kg (100 lb) of commercial-grade, sulfuric acid. Soak for 1 to 4 hours, then thoroughly rinse	Method is used for removing grease, soap, oil, and organic materials that are not entirely surface-clogging substances.[11]
Caustic soaking of ceramic-tube diffusers	Boil tubes in a 10 to 20% caustic soda (lye) solution for 1 hour, then rinse in boiling water several times	Method for removing grease and organic materials.[11]

Table 7.1.   (Continued)

Cleaning Agent Diffuser Type	Procedure	Results and Observations
Detergent cleaning of saran-wrapped, tube diffusers	Tubes are soaked overnight in a detergent solution, then rinsed in a high-pressure jet cleaner. If necessary, soak tubes in a 1:1 solution of muriatic acid and water for 5 to 10 minutes, followed by rinsing	Method for general cleaning of saran-wrapped, tube diffusers. Acid-soaking step is used to remove any remaining inorganic scale.[11]
Cleaning of sock-type diffusers	Remove socks and wash in a household-type washing machine. First rinse, then wash in detergent followed by a second rinsing. If necessary, soak socks in 1:1 solution of muriatic acid and water, then thoroughly rinse	Method for general cleaning of sock type diffusers. Acid soaking is used to remove iron or other inorganic deposits. Cleaned socks should look new, if not, repeat entire cleaning cycle.[11]

Table 7.2.   Cleaning methods for fine bubble diffusers in-place.

Cleaning Agent Diffuser Type	Procedure	Results and Observations
Acid treatment of ceramic plates at the Chicago North Side plant, process interrupted	A 50% nitric acid solution, about 250 mL (8 oz) per plate, was used to treat the plates.	Wet-pressure loss was reduced from 0.87 psig to 0.43 psig at 2 cfm/plate. The effects of cleaning lasted about 6 months.[10]
	Concentrated sulfuric acid containing 2% sodium dichromate, about 100 mL/plate, was used in another tank. Two doses applied 43 hours apart were used.	Wet-pressure loss was reduced from 0.83 psig to 0.54 psig. The effects of cleaning lasted about 6 months. Most plates had not been cleaned in over 20 years.
Acid treatment of ceramic plates at the Chicago Calumet plant, process interrupted	The aeration tank was drained and plates were hosed off with a water spray. 50 to 100 mL of an acid solution (either 50% sulfuric acid, 2% sodium dichromate and 48% water, or 80% sulfuric acid, 2% sodium dichromate and 18% water) was applied to the center of each plate, then spread with a rubber squeegee. The acid application was repeated one or more times as needed, then allowed to stand for 2 days. The tank was filled and the air turned on.	Plates were clogged by iron deposits and by dirty air (2.9 mg/1000 cu ft). Frequency of cleaning was 2 to 3 times per year. The 50% solution of sulfuric acid was used initially but later replaced by the stronger solution. The initial wet-pressure loss after cleaning was equivalent to new plates but clogging was faster on the acid cleaned plates.[4]
Acid treatment of ceramic plates at the Chicago Southwest Plant, process interrupted	Concentrated nitric acid, concentrated hydrochloric acid, sodium metaphosphate solution, and a warm 4.7% sodium dichromate, 66.4% sulfuric acid solution were tested. Additional testing with cold sulfuric acid-dichromate solution used two applications of the chromic solution followed by one application of the concentrated sulfuric acid.	The sulfuric acid-dichromate solution was most effective in restoring permeability. The 3-step procedure was used to clean 23000 plates in 1941. The effects of cleaning lasted about 2 months. The acid-cleaning results were improved by admitting city water into the air main and forming the air-water mixture through the plates to wash out the acid residue.[4,10]

Table 7.2.   (Continued)

Cleaning Agent Diffuser Type	Procedure	Results and Observations
Acid treatment of ceramic plates at the Lima, Ohio, plant, process interrupted	Plates treated by an application of weak muriatic acid.	Plates were clogged by hard-water scale.[12]
Acid cleaning of ceramic plates at the Milwaukee Jones Island plant, process interrupted	Tank was drained, hosed down, filled with several inches of service water, manually wire brushed, hosed down and drained. After drying, a 5:1 inhibited muriatic acid and water solution was sprayed onto the plates using a hand-held, lawn-type sprayer. After 20 to 30 minutes of soaking, the air was turned on to purge the plates and the plates were washed with high-pressure, water spray.	The plates were clogged with iron deposits from use of spent pickle liquor added to the aeration tanks for phosphorus precipitation. The cleaning procedure significantly reduced the wet pressure and improved air distribution. Bubble release pressure (BRP) data indicated that cleaning was not complete or uniform. Average BRP data on the dirty plates ranged from 5.6 to 81.3 inches of water and the coefficients of variation were between 0.30 and 1.31. After cleaning average BRP data were between 6.3 and 30.2 inches of water. The coefficients of variation were between 6.3 and 30.2 inches of water. The coefficients of variation were from 0.20 to 1.03.[13]
Caustic soda cleaning of ceramic plates at the Wards Island, N.Y., plant, process interrupted	Tanks were drained, hosed down, brushed, allowed to dry, treated with a solution containing 25% caustic soda, and allowed to stand for 24 hours. The caustic treatment was repeated in the first pass, then the tank was put back in services without rinsing.	Clogging was caused by organic growths and deposits. The plates were cleaned every 1 to 2 years.
Sandblasting of ceramic plates at Peoria, Ill.	Plates were sandblasted top and bottom.	Wet pressure drop was reduced 1.3 psig on plates in service 15 years. 5 years later wet pressure was reduced by 1.9 psig after sandblasting.[4]
Sandblasting of ceramic plates at the Milwaukee Jones Island plant, process interrupted	Plates were sandblasted top only.	MOP-5 (1952) reported good results with sandblasting at Milwaukee for removing iron deposits. Studies conducted in 1980 showed sandblasting to be detrimental and the more effective acid-cleaning procedure presented above was developed.[4,14]
Acid gas cleaning of ceramic disc diffusers at Seymour, Wis., no process interruption	Air-flow rate was increased to approximately 3 cfm/diffuser, HCl gas was injected into the air stream and dynamic-wet pressure was monitored at 5 to 15 minute intervals. The cleaning process was stopped when no appreciable reduction in it was noted. The gas lines were then purged with nitrogen gas and air-flow rates returned to normal. (Process patented by Water Pollution Control Corporation, Milwaukee, Wis.)	Gas cleaning was effective for recovering the dynamic wet pressure of fouled diffusers. Cleaning took 1 to 2 hours and used 0.1 to 0.15 lb of HCl per diffuser. Estimated cleaning frequency was 3 times per year. Air bumping (doubling the air flow for ½ hour) twice per week was effective in reducing the frequency of acid gas cleaning.[15]

Table 7.2. (Continued)

Cleaning Agent Diffuser Type	Procedure	Results and Observations
Steam cleaning of ceramic dome diffusers at the Madison, Wis., Nine Springs plant, process interrupted	Standard steam cleaning procedure applied to surface of diffusers.	Organic slimes were effectively removed using the steam cleaning. Air distribution and oxygen transfer seemed to improve dramatically after cleaning.[16]
High pressure hosing of ceramic domes and plates at Madison, Wis., process interrupted	High pressure hosed in tanks routinely once per year.	Successfully maintained high efficiencies for over 8 years with domes.[16]
Acid cleaning of plastic-disc diffusers, no process interruption	Formic acid is injected into the process air stream.	Pilot-scale testing and full-scale experiments were successful in cleaning the plugged diffusers.[16]
Acid gas cleaning of ceramic disc diffusers at Lakewood, Ohio, no process interruption	See Seymour, Wis., cleaning procedure. 330 diffusers were cleaned using 960 cfm of air and 31 pounds of HCl gas over a 57-minute cleaning period.	330 diffusers were cleaned. Fouling was caused by frequent process air interruptions during blower rehabilitation work. Dynamic wet pressure was reduced from about 12 inches of water to about 6 inches versus 5 inches for new discs.[8]

observed. Although the surface patterns generated by different types of aeration systems and configurations will vary, the important thing is to become familiar with the normal operation of the system so unusual circumstances can be detected.

Generally, keeping a relatively uniform DO concentration throughout the aeration basin is desirable. If the basin operates as a series of complete-mix cells, the aeration intensity per cell may have to be adjusted to match variations in the demand as the wastewater becomes stabilized passing from one cell to the next.

Operation of sparged aerators requires proper maintenance and control of the air flow to each unit. If only one aerator is operated per blower, excessive air flow to the unit resulting in impeller flooding must be avoided. If multiple units are operated off a central air-supply manifold, the air flow must be properly distributed to each unit. Unbalanced, wandering, or excessive air flow beyond the design limit may lead to impeller flooding. An air-flow meter installed on each aerator air-feed line is recommended for proper air-flow balancing.

SHUTDOWN

If an aeration basin must stand idle for more than 2 weeks, it should be drained and thoroughly cleaned. When taking an aeration basin out of service, these steps should be followed:

1. Stop the wastewater and return sludge flow to the basin.
2. Just before opening the drain lines and starting the drain pumps (if necessary), turn off the aerator drives. This will avoid excessive splashing, extreme vortexing, or surging that could damage the equipment. Continue to feed air to spargers until the liquid level is below the sparging device so that solids do not enter the air piping. Monitor and adjust the air-flow rate as the liquid level falls.
3. Once the basin is drained, check to make sure that the groundwater pressure-relief valves are operational so uplift pressures do not cause damage.

4. To avoid odor problems, wash down the basin walls and floor, the aerator-wetted parts, and any air piping.
5. Inspect the aerator shafting and impeller and make repairs as needed. Materials accumulated on shafts and impellers should be removed so that the unit is ready to be put back into service.

Units that have been in service but will be idle for less than 3 months should be run weekly for 10 to 15 minutes to keep the gears and bearings coated with oil. This will protect the gears from rusting if condensation forms when temperature changes occur.[21]

Units that will be inactive for more than 3 months should be prepared for storage following the motor- and gear-reducer, manufacturer's recommendations. Special care should be taken in cold climates to prevent frost damage to the tank structures.

PREVENTIVE MAINTENANCE

The preventive maintenance program should include the major aeration equipment—motors and gear reducers.

Motors. To ensure continued reliable operation of an electric motor, the motor should be kept clean and properly lubricated. Motors should be inspected at regular intervals with the frequency depending on the type of motor and the service.

Windings should be cleaned by blowing or vacuuming dust from them. Ventilation openings must be kept free of dirt. If dust and dirt are to be removed with a vacuum cleaner, nozzles should be nonmetallic. Deposits of dirt and grease may be removed by using a commercially available, low-volatile solvent.

Terminal connections and assembly hardware may loosen from vibration during service. They should be checked and tightened, if needed.

Insulator resistence should be checked at normal operating temperatures and humidity conditions to determine possible deterioration of insulation. If wide variations in resistence are detected, the motor should be reconditioned.

Motor bearings will require periodic lubrication. The manufacturer's recommendations on frequency, procedure, and lubricant type should be followed. In general, larger horsepower motors will require more frequent lubrication.

Gear Reducers. To maximize the life of gear reducers, the manufacturer's recommended lubrication for the gears and bearings should be followed. The gear case must be filled to the proper level with an oil of the correct viscosity for the operating conditions encountered. In areas with wide seasonal temperature changes, seasonal oil changes will be necessary. Changing oil viscosity in October and April is common.

The recommended oil-change frequency will depend on the service conditions, the gear type, and the lubricant used. For units operating under favorable conditions and using petroleum-based lubricants, the normal oil-change frequency is about 2500 hours or 6 months, whichever occurs first. More frequent oil changes, at intervals of 1 to 3 months, are necessary if the gear drives operate under conditions that tend to deteriorate oil or cause condensation. Under some service conditions, certain gear types using synthetic lubricants may only require lubricant changes if the reducer must be drained for other maintenance. Lubricant suppliers and independent laboratories can provide an analysis of an oil sample to determine if an oil change is required.

Various types of bearings are used in gear reducers. The manufacturer's literature should be studied for instructions on how to lubricate the bearings. In general, the frequency of re-lubrication depends on the extent of atmospheric contamination, excessive moisture, variations in ambient temperatures, and actual bearing operating temperature. The lubrication frequency can vary from 1 to 6 months. Bearings operating at higher temperatures require fresh grease at more frequent intervals.

TROUBLESHOOTING

Mechanical problems associated with aerator drives include noisy operation, abnormal heating, oil leaking, noisy bearings, overheated bearings, and vibration. Many of these problems are related to improper lubrication. The manufacturer's literature should be referenced for detailed troubleshooting guidelines.

Of all the mechanical problems that can be encountered, vibration is probably the most misunderstood phenomenon. Any amount of vibration is viewed as a sign of impending disaster. Although vibration is not a good sign, it is inherent in any rotating machine. In recent years, there has been a large increase in the use of vibration-measuring equipment as part of a general maintenance program. Vibration cannot be measured *per se.* What is actually measured is the peak-to-peak displacement at a given frequency. The frequency of the displacement must also be measured to make the data useful. A vibration "signature" can be taken of a piece of equipment when it is new and saved for reference. Additional signatures taken over time can give a warning of impending bearing failure or mechanical looseness.

The signature analysis will exhibit peaks at certain frequencies. The frequencies may correspond to shaft rpm, ball-passing and gear-tooth frequencies, and harmonics of these. By denoting at which frequency a vibration increase has occurred, it is possible to pinpoint the problem. Because of their slow speed of operation and large overhung load, slow-speed aerators are designed to take more vibration than other types of rotating equipment. Assistance in performing vibration tests and data analysis can be provided by some aerator manufacturers. Independent-testing firms are also available.

Even if an aerator is performing well mechanically, process-related problems may arise. In installations where more than one aerator is installed in an aeration basin, hydraulic interactions between the aerators can occur resulting in loss of OTE and increased power usage. Unacceptable hydraulic interactions, also known as surging, are usually controlled by baffling in the aeration basin. The adequacy of the baffling can be tested by monitoring the power drawn by the aerator over time.

Mechanical aerator efficiency and oxygen transfer capability can also be greatly affected by changes in the impeller characteristics caused by fouling and wear.[23] If the process-monitoring program indicates a loss in transfer efficiency accompanied by an increase in power draw, impeller fouling should be checked. If the loss in transfer efficiency is accompanied by no change in power draw, the problem is usually caused by impeller wear.

REFERENCES

1. "Plant Maintenance Program." Manual of Practice OM-3, Water Pollution Control Federation, Alexandria, VA (1982).
2. Danly, W.B., "Biological Fouling of Fine Bubble Diffusers." M.S. Thesis, Civil and Environmental Engineering, University of Wisconsin-Madison, WI (1984).
3. American Society of Civil Engineers, "Measurement of Oxygen Transfer in Clean Water." ASCE Standard, Am. Soc. of Civ. Eng., New York, N.Y. (1984).

4. Redmon, D.T., *et al.* "Oxygen Transfer Efficiency Measurements in Mixed Liquor Using Off-Gas Techniques." *J. Water Pollut. Control Fed.*, **55,** 11, 1338 (1983).

5. Marx, J.J., and Schroedel, R.B., "Development and Implementation of a Full-Scale Aeration System Test Program." Paper pres. 59th Annual Meeting of the Central States Water Pollution Control Association, Milwaukee, WI (1986).

6. "Air Diffusion and Sewage Works." Manual of Practice No. 5, Water Pollut. Control Fed., Alexandria, VA (1952).

7. Wirts, J.J., "Cleaning Air Diffusion Media." *Water and Sewage Works*, **94,** R-201 (1947).

8. Winkler, W.W., "Fine Bubble Ceramic Diffuser Maintenance." Paper pres. Annual Meeting of the New England Water Pollution Control Association, Boston, MA (1984).

9. Schade, W.F. and Wirts, J.J., "Diffuser Plate Cleaning Versus Compressed Air Cost." *Sew. Works J.*, **14,** 1, 81 (1942).

10. Beck, A.J., "Diffuser Plate Studies." *Sew. Works J.*, **8,** 22 (1936).

11. Blodgett, J.H., "Air Diffusion with Saran Wrapped Tubes." *Sew. and Ind. Wastes*, **22,** 10, 1290 (1950).

12. "Aeration in Wastewater Treatment." Manual of Practice No. 5, Water Pollut. Control Fed., Alexandria, VA (1971).

13. Wisely, W.H., "Summary of Experiences in Diffused Air Activated Sludge Plant Operation." *Sew. Works J.*, **15,** 909 (1945).

14. "Evaluation of Diffuser Cleaning Methods Employed at the Jones Island Wastewater Treatment Plant." Rept. prepared for the Milwaukee Metropolitan Sewerage District, Ewing Engineering Company, Milwaukee, WI (1982).

15. "Restoration of the East Plant Aeration tank No. 11 at the Jones Island Wastewater Treatment Plant." Rept. prepared by Milwaukee Metropolitan Sewerage District, Milwaukee, WI (1980).

16. Vik, T.E., *et al.*, "Full-Scale Operating Experience Utilizing Fine Bubble Ceramic Aeration with In Place Gas Cleaning at Seymour, Wisconsin." Paper pres. 57th Annual Meeting Central States Water Pollution Control Association, Minneapolis, MN (1984).

17. Bretscher, V., and Hager, W.H., "The Cleaning of Wastewater Aerators." *Water and Wastewater Eng.*, **6** (1983).

18. Boyle, W.C., and Redmon, D.T., "Biological Fouling of Fine Bubble Diffusers." ASCE Environ. Eng. Div., *J. Environ. Eng.*, **109,** 5 (1983).

19. "Some Cleaning Techniques for Fine Bubble Dome and Disk Aeration Systems." Rept. prepared by County Sanitation Districts of Los Angeles County, Los Angeles, CA (1984).

20. "Aeration System Evaluation Study—Interim Report No. 2." Rept. Prep. for the Green Bay Metropolitan Sewerage District, Donohue and Associates, Inc., Greenbay, WI (1987).

21. Mixing Equipment Company, Inc., "A Practical Guide to Mixer Maintenance." 2nd Edition, Mixing Equipment Co., (1985).

Chapter 8
Specifying and Testing Aeration Systems

*I*NTRODUCTION

Historically, many methods have been used to test and specify aeration equipment. The varied methodologies, at times, have led to confusion and misrepresentation of aeration equipment performance. Furthermore, equipment suppliers, consultants, and users often do not use the same nomenclature when they report aeration-equipment capabilities.

Aeration equipment specification is a vital part of selecting efficient and cost-effective aeration devices for use in biological and physical-chemical treatment systems. To enable an equipment supplier to properly specify and quote aeration units, the prospective user or his design engineer must provide accurate and detailed field-requirement information. The supplier can then reply to user requests by providing equipment performance based on reliable clean-water test data, gathered through standard techniques and sound judgment using experience from previous field applications of his equipment. In short, uniform specifications will minimize user misunderstanding of equipment quality and performance so that realistic evaluations and comparisons of equipment can be made.

Reliable clean-water data are only the first step in understanding aeration-system capabilities. Measuring oxygen transfer in the field with aeration equipment operating under actual process conditions is also imperative. Various field or respiring system test methods are available and must be used with care depending on the specific-test facilities. With the increased use of fine-pore, diffused-aeration systems in the field over the past few years, the off-gas testing method has been improved and used more extensively for diffused-aeration devices. Although none of the field methods should be labeled as standard, the unsteady-state procedures (with and without hydrogen peroxide addition) in either continuous or batch mode are considered to be among the most cost-effective approaches for field testing of aeration devices.

The following sections discuss specification and testing protocols that will enhance understanding in measuring aeration-equipment performance; the subsequent transactions between vendors, consultants, and equipment users are simplified.

AERATION EQUIPMENT CONSIDERATIONS

Before equipment selections can be made, field requirements and vendor specifications must be determined.

FIELD REQUIREMENTS FROM USER

The first step in proper equipment selection for an aeration application is for the prospective user to establish the aeration-system field requirements. The important elements defining the field requirements are described in the following list:

1. Site location. Elevation (above sea level): ambient high summer temperature, ambient low winter temperature (°C);

2. Aeration volume. Process water depth, tank and basin configuration;

3. Oxygen demand. Minimum, average, maximum;

4. Mixing requirement. Capability to uniformly suspend the MLSS/MLVSS concentrations specified;

5. Operating conditions. Process water temperature (°C): minimum, average, maximum (absolute allowable level to protect biological organisms);

6. Process water transfer characteristics. Range of alpha (α) and beta (β) (Note: beta is related strictly to the impurities in the process water, whereas alpha, which may vary with time and location in the aeration volume, is related to the process-water characteristics, aeration device, and aeration volume design configuration, complete mix versus plug flow);[1]

7. Residual DO (mg/L). Minimum (5 percentile), average, maximum (95 percentile); probability levels as discussed below;

8. MLSS/MLVSS concentrations (mg/L). Minimum, average, maximum; and

9. Aeration equipment. Desired type of aeration system [mechanical, diffused (coarse or fine pore), or mechanical-diffused], required construction materials, required performance testing and necessary quality control for installation.

One beneficial method of reporting oxygen demand (item no. 3 in the preceding list) is based on occurrence probability (in percent of time a specific value is likely to be lower than a given point in a statistical distribution of the data). For example, rather than giving an absolute minimum, a value should be specified at which loadings would not be low more than 5% of the time (5% probability level). On the other hand, rather than maximum loadings, a value should be specified at which the loading is not likely to be exceeded more than 5% of the time (95% probability level). If available, this statistical information can provide a more effective basis for aeration design and system-control flexibility. Oxygen-demand distribution through aeration tank and basin should also be specified for plug-flow and step-feed systems so that the manufacturer can specify proper distribution of oxygen-transfer equipment throughout the aeration volume.

The importance of each of the previously listed field requirements in proper aeration-system selection has been discussed in the previous chapters. In summary, the combination of requirements as specified by the user will provide the equipment suppliers adequate information to specify and quote a site-specific, aeration system.

EQUIPMENT SUPPLIER SPECIFICATIONS AND QUOTES

Equipment suppliers should provide users with detailed mechanical and structural and performance characteristics of their equipment, including reliable clean-water performance data. Clean water data must be reported at standard conditions and be accompanied by a description of the test conditions under which the data were derived. This information will allow the prospective user to judge the appropriateness of clean-water test data for specific aeration application. Because clean-water transfer data are the primary basis for specification of aeration equipment, this basic data must be gathered in accordance with the protocols outlined in the ASCE Oxygen Transfer Testing Standard[2] summarized in a following section of this chapter.

The prospective user or the design engineer should transpose the clean-water performance with performance under the field conditions noted above for the particular application. This transposition will involve the use of process-water oxygen demands, alpha, temperature, residual DOs, and beta. This procedure for standard clean water OTR adjustment to process conditions was discussed in Chapter 3 and in a later section of this chapter entitled "Clean Water Testing." The importance of appropriate alpha level use cannot be overstated, and the design engineer should exercise good judgment based on the fundamental knowledge of aeration, the specific equipment, and the past experience in the field. The equipment supplier, based on experience with the different equipment applications, is responsible for reviewing the appropriateness of the alpha values for the specific application. Feedback by the supplier to the design engineer is essential to selection of the appropriate design basis.

When the above adjustments and instructions are completed, the design engineer can select the appropriate number of aeration units to satisfy design criteria. For a complete supplier response to customer inquiries, the following information should accompany the specification and quote for aeration equipment:

- aeration unit number required to meet critical design conditions, including carrer consideration of adequate turn down capability;
- total aeration unit cost;
- delivered power required to drive the aeration units, if appropriate;
- total air required by aeration units, if appropriate;
- the air-distribution system design, including headloss calculation, maximum pressure, and pipe and orifice size(s), if appropriate;
- delivered power required to the air blower, if appropriate;
- inlet air filtration requirements, if appropriate;
- clean-water data, along with standard test conditions, used as basis for SOTR at minimum, average, and maximum oxygen demands;
- calculations to transform SOTR to OTR_f;
- equipment construction materials including detailed drawings and specifications that outline device mechanical and structural integrity;
- quality assurance/quality control (QA/QC) programs used in equipment manufacture, shipping, storage, and installation; and
- full-scale testing program to demonstrate equipment performance guarantees (aeration and mixing, at minimum, average, and maximum oxygen demand periods, if practical).

CLEAN WATER TESTING

SIGNIFICANCE IN THE ENGINEERING FIELD

Many types of aeration devices are used to supply oxygen to a variety of biological and physical and chemical, treatment plants throughout the world. In North America alone, approximately 1.75×10^6 hp (1.3×10^6 kW) is currently used to drive aeration equipment.[3] Because of enormous power consumption and a wide disparity in performance between various aeration-equipment types, accurate aeration-equipment performance knowledge is required. Accurate OTR measurement for aeration equipment is easily obtained under clean-water conditions. Therefore, clean-water test data are the foundation of equipment performance claims; they have become the most common basis for widespread relative-aeration performance comparison for the many commercially available devices. This is not to say, however, that similar performance by two aeration devices in clean water translates to similar performance under process conditions because, as experience in the field has shown, they may be significantly different.

The following section summarizes the standard developed by the ASCE Oxygen Transfer Standards Subcommittee for testing aeration-equipment performance under standard clean-water conditions.[2] This standard is intended to be used by engineers in the preparation of specifications for equipment-compliance testing and by manufacturers in the development of equipment-performance information.

It should be emphasized that clean water testing is not the only performance testing method although it is clearly the most common. Performance testing under process conditions, though sometimes desirable, is typically more difficult, and test results are not viewed as being as accurate as clean water results. Despite these drawbacks, process water testing methods are valuable and will be discussed later.

OTR MEASUREMENT

This method covers OTR measurement as an oxygen mass per unit dissolved in water by an oxygen transfer system that operates under a given aeration rate and power conditions. The method is applicable to laboratory-scale oxygenation devices with water volumes of a few liters as well as to full-scale systems with water volumes typical of those found in the activated sludge wastewater-treatment process. The procedure is valid for many different mixing conditions and process configurations.

The primary test result under specified aeration rate and power conditions is expressed as the standard OTR, or SOTR, a hypothetical oxygen mass transferred per unit of time at zero DO concentration, 20°C water temperature, and 1.00 atmosphere barometric pressure. The method is intended primarily for clean water meeting the requirements specified in the "ASCE Standard." The results, however, can be applied to estimate OTR under process-water conditions.

The test method is based on DO removal from the water volume by sodium sulfite in the presence of a cobalt catalyst followed by reoxygenation or reaeration to almost the saturation level. The water volume DO inventory is monitored during the reaeration period by measuring DO concentrations at several determination points selected to best represent tank contents. These DO concentrations may be either sensed *in-situ* with membrane probes or measured by the Winkler or probe method in conjunction with pumped samples. The method specifies a minimum number, distribution, and range of DO measurements at each determination point.

The data obtained at each determination point are then analyzed by a simplified mass-transfer model (Chapter 3) to estimate apparent volumetric mass transfer coefficient, K_La, and equilibrium spatial average DO saturation concentration, C_∞^*. The basic model is given by:

$$C = C_\infty^* - (C_\infty^* - C_o) \exp(-K_L a t) \qquad (1)$$

Where:

- C = Average spatial DO concentration, mg/L;
- C_∞^* = Spatial average steady state DO saturation concentration, the concentration attained as time approaches infinity, mg/L;
- C_o = Average spatial DO concentration at time zero, mg/L; and
- K_La = Apparent spatial average volumetric mass transfer coefficient, t^{-1}, defined so that;

$$K_La = \frac{\text{rate of mass transfer per unit volume}}{C_\infty^* - C}$$

Nonlinear regression is used to fit Equation 1 to the DO profile measured at each determination point during reoxygenation. In this way, estimates of K_La and C_∞^* are obtained at each determination point. These estimates are adjusted to standard conditions and the SOTR is obtained as the product average of the adjusted point K_La values, the corresponding adjusted point C_∞^* values, and the tank volume.

KEY ELEMENTS

The key elements and foundation of the OTR measurement test are the definition of terms used during aeration testing, subsequent data analysis, and final result reporting. A consistent nomenclature has been established with more logical and understandable terminology than the numerous and varied symbols used historically. When consistently used in the field, this standard nomenclature will eliminate much of the difficulty in interpretation of aeration literature.

Procedure and data analysis. Actual test details that include specifications on test preparation, test set-up (i.e., tank geometry and aeration placement), minimum number of tests required, and itemized-test procedures are outlined in the reference.[2] Chemical addition methodologies, number and location of DO sample points, and discussion of DO measurements are provided. Once the clean-water data have been collected, they must be analyzed to yield corresponding OTR results. Procedures are given for preparation of the data for a statistically sound analysis that uses nonlinear regression. The aeration model (Equation 1) uses DO versus time data provided by the clean-water test. This analysis will provide the best estimates for the three model parameters, K_La, C_∞^*, and C_o, so that the model best describes the variation in DO with time at each determination point during the test.

RESULTS INTERPRETATION AND REPORTING

The K_La and C_∞^* results from nonlinear-regression analysis are then checked for consistency of different determination points in a test tank and for replicate tests. Criteria are provided for spatial-variation judgment in the tank and the adequacy of the number and location of determination points. For example, with a minimum of four determination points in a test tank, variation of average K_La or C_o^* point values should be limited so that three-fourths of the values are within $\pm 10\%$ of the mean value for the tank. If spatial variations are greater than this value, consideration should be given either to using a greater number of determination points or testing by tank sections. Testing by tank sections would be mandatory for a plug-flow, aeration volume. Considering K_La_{20} results from three or more replicate tests, the determination point values in each replicate should not vary by more than $\pm 15\%$ from the mean point value. Replicate test results that exhibit greater variation are judged invalid and should not be used for measured SOTR calculation.

Following the check for spatial uniformity and reproducibility, the transfer rates must be converted to SOTR. The standard temperature and pressure conditions for SOTR are usually 20°C and 1.00 atmosphere. The SOTR, however, is a hypothetical value. It is based on zero DO in the oxygenation zone, which is not usually a desirable condition in real oxygenation systems treating process water. The average SOTR value is calculated by averaging the values at each of the n determination points:

$$SOTR = V \sum_{i=1}^{n} \frac{K_La_{20i}C^*_{\infty\,20i}}{n} \tag{2}$$

$$= \frac{\sum_{i=1}^{n} SOTR_i}{n}$$

from Equation 13, Chapter 3.

Where:

$$SOTR_i = K_La_{20i}C^*_{\infty 20i} V \tag{3}$$

V = Volume of water in the test tank (mil gal)

Frequently, the standard aeration efficiency (SAE) or rate of oxygen transfer per unit power input is of interest and can be calculated from:

$$SAE = SOTR/\text{power input} \tag{4}$$

where SAE is normally expressed in kg/kW-hr (lb/hp-hr); and power input may be expressed as delivered, brake, wire, or total wire power and must be specified.

For diffused-aeration systems, the term standard oxygen transfer efficiency (SOTE), which refers to the fraction of oxygen transferred from an injected gas stream under standard conditions, is of interest. SOTE may be calculated for a given flow rate of air by:

$$SOTE = SOTR/W_{O_2} \tag{5}$$

where SOTE is expressed as a fraction and W_{O_2} is mass flow rate of oxygen, lb/hr.

In all cases, the fundamental clean water performance measure is SOTR. The measure of SAE and SOTE are supplemental aerator-performance expressions.

In reporting test results, the key items to be included are: test purpose; test-site description including site elevation; test-tank details; water source and quality (total solids, surface active agents, oil and grease test conditions including air temperature; barometric pressure, water temperature, depth, and volume; cobalt and sodium sulfite use; monitoring point locations and measured power and air-flow rates (if appropriate); test results including K_La, K_La_{20}, C^*_{∞}, $C^*_{\infty 20}$, and d_e (effective depth) for each monitoring point and average for replicates; and SOTR (SAE and SOTE if appropriate) for each replicate test. Reporting under this standard format will enhance the understanding and value of the clean-water test results.

TESTING UNDER PROCESS CONDITIONS

BACKGROUND

Once aeration equipment has been installed and is operating under process conditions, field performance should be examined relative to design estimates. Several conventional and innovative methods are available to test the aeration equipment during process operation. All of these techniques are referred to as respiring-system tests.

In general, the methods can be categorized according to the rate of DO change in a given reactor (or reactor segment). Systems in which the DO change rate is zero at any given point are referred to as steady-state systems; the others are classified as non steady-state systems. Some of the methods require direct oxygen-uptake rate measurement; others do not. In some cases, influent wastewater may be diverted from a reactor being tested. These are referred to as batch tests. The term "continuous test" is used for those cases in which the influent waste-water flow is not diverted.

Many respiring-system test methods do not require a direct measure of the oxygen-uptake rate. These have been broadly categorized as the mass balance method, the off-gas method, the inert-gas tracer method, and the non steady-state methods. The mass balance method requires data on the net change in the waste-oxidation level between all influent and effluent liquid flows. The off-gas method is a mass balance on oxygen that includes both the liquid and gas streams. The inert-tracer method indirectly measures the oxygen-transfer rate by determining the transfer rate of a radioactive or stable inert gas tracer. For the non steady-state methods, the reactor DO level is adjusted at the beginning of the test to be either greater than or less than the steady-state DO. These tests are referred to as "Non steady-State Batch Tests" if the influent-wastewater flow is discontinued for the test and "Non steady-State Continuous Tests" if the influent wastewater flow is continued during testing. Oxygen-uptake rate is not required for data analysis in any of these tests but is often measured to ensure that relatively constant-operating conditions prevail during the test.

The other commonly used methods do require direct measurement of the oxygen-uptake rate, R, of a respiring biological system. These methods, which are carried out with little or no variation in DO in either batch or continuous-flow systems, are referred to as "Steady-State Batch Tests" and "Steady-State Continuous Tests," respectively.

The relationship between the steady-state and non steady-state tests (either batch or continuous) may be readily seen from a hypothetical example (Figure 8.1). In Figure 8.1, the oxygen concentration (designated as C) at one point in a reactor is shown to decrease from C_{R1} to C_{min} after the aeration equipment is turned down; then it is shown to increase from C_{min} through C_o to C_{R2} after the aerators are turned back up (C_{R1} may or may not be equal to C_{R2}). The concentration C_o simply designates the DO concentration corresponding to time t_o—the time at which the investigator decides to initiate the full-data analyses spectrum. The DO concentration obviously increases whenever the oxygen supply exceeds the biological consumption rate, remains constant when the two rates are equal (in this case, at C_{R1} or C_{R2}), and decreases when the consumption rate exceeds the supply rate. In clean-water testing, sodium sulfite is used to reduce the DO level. In clean water, however, the rate of oxygen consumption is zero after time t_o, with the test proceeding to equilibrium at the effective system saturation concentration, C_∞^*. In a respiring system, the steady-state oxygen concentration, C_{R2}, is reached at time t_o plus t_R. The effective saturation con-

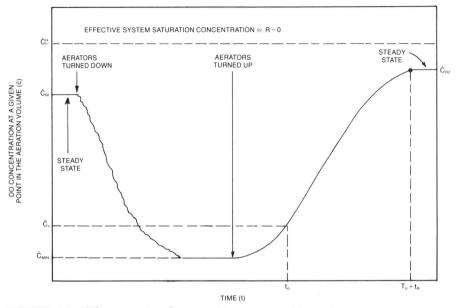

FIGURE 8.1—DO versus time for a respiring system with varying aeration conditions.

centration of oxygen that would result in the mixed liquor for a zero-respiration rate is given in Figure 8.1 as $C_{f\infty}^{*}$.

Material balances have been developed for continuous flow and batch reactors to both properly analyze the data generated from many of the test methods and to understand those assumptions that are required for data-handling ease. General schematics for both continuous-flow and batch systems are presented in Figure 8.2. The term "batch" refers to the liquid phase only. Batch analysis allows for sludge solids recycling but requires that there be no net liquid flow.

In Figure 8.2, the symbols used represent the following:

- C, S, X, and N represent concentrations in the liquid phase of oxygen, substrate, biological organisms, and ammonia nitrogen, respectively.
- C_R will be used to designate the steady-state concentration of oxygen at a biological uptake rate of R.

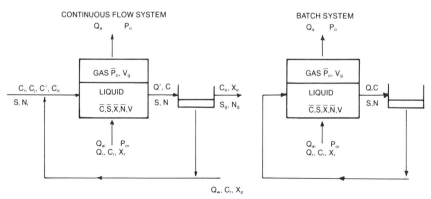

FIGURE 8.2—General schematics and nomenclature for continuous flow and batch respiring systems.

- Q and Q_a represent flow rates for liquid streams and gas streams, respectively. Q_w represents the flow rate for waste-activated sludge.
- P_o and P_{oi} represent the partial pressures of oxygen in the gas phase in the reactor and in the inlet, respectively.
- V and V_g represent the reactor volumes of the liquid and gas phases, respectively.
- The subscripts i, o, r, and e in the liquid phase are used to designate the influent wastewater, the combined streams entering the reactor, the recycle sludge flow, and the secondary clarifier effluent, respectively. No subscripts are used for either reactor concentrations or reactor effluent concentrations. A bar (for example $\overline{C}$) is used to represent spatially averaged concentrations of all reactor concentrations. Note finally that the symbol Q' is used to designate the liquid flow entering and leaving the continuous-flow reactor (instead of Q_o). Q_o has been defined in a previous section.

The general mass balance on oxygen for the liquid phase is given in Equation 6 for a continuous-flow reactor and in Equation 7 for a batch reactor with C at t = 0 equal to C_o for both equations:

$$V \left[\frac{d\overline{C}}{dt} \right] = Q'(C'_o - C) - VR + VK_L a_f(C^*_{\infty f} - C) \quad (6)$$

$$V \left[\frac{d\overline{C}}{dt} \right] = Q_r(C_r - C) - VR + VK_L a_f(C^*_{\infty f} - C) \quad (7)$$

For completely mixed reactors, $\overline{C}$ is equal to C.

The equations used to analyze results from field tests for steady-state and non steady-state conditions result directly from Equations 6 and 7. The assumptions required to develop these equations are described in Tables 8.1 and 8.2 for both batch and continuous tests. Many of the assumptions, especially for the continuous-flow systems, may not be valid in the field test. The validity of the assumptions must be evaluated with each test application. A detailed discussion of this issue is contained in an EPA report titled "Development of Standard Practices for Evaluating Oxygen Transfer Devices."[4]

Table 8.1   Assumptions necessary to develop equations for steady- and non steady-state batch tests.

Assumptions	Test Conditions
1. Aeration volume DO (C_R) is constant and the reactor contents are completely (uniformly) mixed	Steady-state conditions have been achieved and maintained; they are required for non steady-state test.
2. Recycle sludge flow (Q_r) is constant zero	Recycle flow rate maintained, constant, or discontinued.
3. Recycle DO (C_r) is constant	Steady operation of recycle system if in use during the test period (for example sludge blanket level constant).
4. Aeration volume oxygen uptake rate (R) is constant	Aeration volume biological solids (X) and recycle sludge flow (Q_r) remain constant; carbonaceous (S) and nitrogenous (N) substrates are near zero during the test (if nitrification occurs in the test system.)
5. Effective oxygen transfer rate ($K_L a_f$) is constant	Carbonaceous substrate (S) is near zero during the test; alpha (α) value remains constant during the test period.

Table 8.2   Assumptions necessary to develop equations for continuous steady- and non-steady-state tests.

Assumptions	Test Conditions
1. Aeration volume DO (C_R) is constant and the reactor contents are completely (uniformly) mixed	Test time is short; they are not required for non-steady-state test
2. Reactor influent flow (Q') is constant	Test time is short; variation in influent wastewater flow (Q') is negligible during test period; recycle sludge flow (Q_r) is held constant.
3. Influent DO (C_o^i) is constant	Test time is short; variations in influent wastewater flow (Q_i) and DO (C_i) are negligible during test period; recycle sludge flow (Q_r) is held constant.
4. Aeration volume oxygen uptake rate (R) is constant	Test time is short; aeration volume biological solids (X) and recycle sludge flow (Q_r) remain constant; influent wastewater flow (A_i) and carbonaceous substrate (S) variations are negligible during test period. (Note that R may be difficult, if not impossible, to accurately determine for systems with high-organic loadings.)
5. Effective oxygen transfer rate (K_La_f) is constant	Test time is short; alpha (α) value remains constant during the test period.

Testing Methods

A comprehensive review of the available test methods for field oxygen transfer measurement is presented in another report[4] and summarized in the following sections.[5-8] A tabulation of those factors on which an appropriate oxygen transfer method can be selected (cost, precision, accuracy, basin configuration and aeration device) is shown in Table 8.3.

Steady-state continuous test. Steady-state continuous testing involves simultaneous measurement of DO and oxygen uptake rates in a full-scale aeration tank with influent wastewater flow to the tank. For this testing mode, the field transfer rate may be calculated using:

$$K_La_f = \frac{R - \dfrac{Q'}{V}(C_o' - C_R)}{C_{\infty f}^* - C_R} \tag{8}$$

Where:
K_La_f	=	field oxygen transfer rate (h^{-1}),
R	=	oxygen uptake rate (mg/L·h),
Q'	=	hydraulic throughput rate (mil gal/h)
V	=	aeration volume (mil gal)
C_o'	=	influent DO level (mg/L),
C_R	=	aeration tank DO level (mg/L), and
$C_{\infty f}^*$	=	effective field DO saturation level (mg/L).

Steady-state batch test. Steady-state batch (or batch-endogenous) testing involves simultaneous measurement of DO and oxygen uptake rates in a full-scale aeration tank without influent wastewater flow to the tank. For this testing approach, the field transfer rate may be calculated using:

$$K_La_f = \frac{R - \dfrac{Q_r}{V}(C_r - C_R)}{C_{\infty f}^* - C_R} \tag{9}$$

Table 8.3 Selected factors affecting oxygen transfer field testing for estimation of K_La_f.

Factors	Oxygen Transfer Tests			
	Steady State	Non-Steady State	Off-Gas	Inert Gas Tracers
Sensitivity To Variations in:				
Qi	—	—	+	+
R	—	—	+	+
α	—	—	+	+
DO	—	—	+	+
qi	—	—	+	—
Accurate Measure of:				
R	—	+	+	+
DO	—	—	+	+
$C^*_{\infty f}$	+	+	—*	+
qi	+	+	—*	+
Other	+	+	**	***
Basin Configuration				
Plug	—	—	+	+
CSTR	+	+	+	+
Aerator Type	+	+	—	+
Costs				
Manpower	+	0	0	0
Analytical	+	+	0	—
Capital Invest.	+	+/0	0	—
Calculations	+	0	0	0
Estimated Precision	—	0	+	+

  * Calculates OTE directly; requires $C^*_{\infty f}$ and q_i to estimate K_La_f.

  ** Requires accurate estimates of CO_2 and water vapor in gas.

*** Requires accurate estimate of K_{tr}/K_La, especially in diffused air systems.

  + positive response (eg., non sensitive, less costly, more precise, and easier)

  0 intermediate response

  — negative response

Where:

    Q_r = Recycle sludge flow (mil gal/h),

    C_r = Recycle flow DO level (mg/L), and

the other terms have been defined previously.

Non steady-state continuous test. Non steady-state continuous testing involves a DO depletion or increase in the test tank by changing power levels of the aeration devices or a DO increase in the test tank by adding hydrogen peroxide (H_2O_2) to the test tank. Changing aeration power levels in the test tank may result in an increase or decrease in DO. If power levels are decreased, biological action (oxygen uptake) reduces DO in the test tank much like what occurs in clean water non steady-state tests when sodium sulfite is added. After the DO level has been sufficiently depressed, aeration is increased or reintroduced to the test tank. By simultaneously monitoring the change in DO level, the process flow, and the oxygen uptake rate (optional), the field OTR may be determined with the use of the classical reaeration approach and the following equation:

$$(C_R - C) = (C_R - C_o)\ \exp - \left[\left(K_La_f + \frac{Q'}{V} \right)t \right] \quad (10)$$

Where:

    C_o = Initial test tank DO level where increased aeration is initiated (mg/L)

On the other hand, if power levels are increased, DO increases and the rate of DO increase is monitored until a new steady state is reached. Similarly, if peroxide is added and is biochemically reduced by the MLSS to oxygen and water, an elevation in test tank DO level to above normal levels will occur. In either approach, by monitoring the DO level decline (deaeration) versus time and the process flow, the field OTR may be determined using the classical reaeration approach and the following equation:

$$(C - C_R) = (C_o - C_R) \exp - \left[\left(K_L a_f + \frac{Q'}{V} \right) t \right] \quad (11)$$

Where:

C_o = Test tank DO level following the peroxide addition and reaction when exponential oxygen-stripping begins and data analysis commences (mg/L)

In the three applications just discussed, the non-linear regression method can and should be used to determine $K_L a$ and C_R.

Non steady-State Batch Test. Non steady-state batch (or batch-endogenous) testing involves (a) a change of DO in the test tank by varying the power level applied to the aeration volume, or (b) increase in DO in the test tank by the addition of hydrogen peroxide *without* influent wastewater flow to the tank. For this test mode, using the same data-collection procedures as in the continuous tests, the field OTRs can be determined by the classical reaeration approach and Equations 10 and 11 by replacing Q′ with Q_r.

Off-gas analysis. Off-gas analysis has been used with full-scale diffused-aeration activated sludge systems and covered tank pure-oxygen activated sludge systems.[6,7,8] A mass balance on oxygen in the gas phase is required; therefore, this method is not applicable to surface-aeration equipment. For unenclosed systems, a collector or hood is placed in the aeration tank's zones of interest so representative samples can be collected and analyzed for residual oxygen. Technique validity depends on accuracy and precision of the various parameters that have to be measured. Obtaining the requisite accuracy and mass balance on the input and discharge gases is difficult under the best of conditions for low OTE (3 to 5%) associated with many diffused-aeration systems. Significant advances, however, have recently been made concerning instruments and techniques for measuring gas-phase oxygen concentration. Testing aeration systems with OTEs of 8% or greater using off-gas analysis is becoming more the rule than the exception. Furthermore, recent efforts to develop off-gas techniques that provide the requisite accuracy and precision in field situations have been reported.[8,11]

OTE may be calculated with off-gas measurements based on the following relationship:

$$OTE = \frac{\rho q_i Y_R - \rho q_o Y_{og}}{\rho_{qi} Y_R} \quad (12)$$

Where

ρ = density of oxygen at temperature and pressure at which gas flow is expressed, M/L^3

q_i, q_o = total gas volume flow rates of inlet and outlet gases, L^3/t, and

Y_r, Y_{og} = mole fractions (or volumetric fractions) of oxygen gas in inlet and outlet gases

In this equation, no estimate of C_∞^* is required but gas flow rates must be accurately monitored for corrections to CO_2 evaluation.

Gas-flow measurements may be omitted from the calculation of OTE by employing molar ratios of inlet and outlet oxygen to the inert-gas fractions as follows:[8]

$$\text{OTE} = \frac{O_2 \text{ in } - O_2 \text{ out}}{O_2 \text{ in}} \qquad (13)$$

$$\text{OTE} = \frac{G_i(M_o/M_i)MR_{o/i} - G_i(M_o/M_i)MR_{og/i}}{G_i(M_o/M_i)MR_{o/i}} \qquad (14)$$

$$\text{OTE} = \frac{MR_{o/i} - MR_{og/i}}{MR_{o/i}} \qquad (15)$$

Where:

G_i	=	mass rate of inerts, (lb/scfm),
M_o	=	molecular weight of oxygen, (lb),
M_i	=	molecular weights of inerts, (lb)
$MR_{o/i}$	=	mole ratio of oxygen to inerts in feed air, and
$MR_{og/i}$	=	mole ratio of oxygen to inerts in off gas.

Tracer technique. A radioactive-tracer technique has been proposed to measure OTRs in any aeration system, either in clear water or in wastewater.[12] The tracer method requires considerable planning, special radioactive counting equipment, and a radioactive materials license to use the radioisotopes, krypton-85, and tritium. The basic concept of the radioactive-tracer technique involves direct mass transfer measurement of krypton-85, which is related to the OTR. The tritium is used to measure tracer dispersion in the aeration tank. The tritium transfer rate, K_r, is assumed to be negligible (i.e., tritium is a conservative tracer). Both tracers are added to the aeration tank at a single point.

When the tracers are dispersed in the aeration tank, the tritium mixes with the mixed liquor while the krypton-85 is stripped off in the gas phase. The key to this procedure is that the krypton-85 stripping rate from the mixed liquor is directly related to the OTR from the gas into the mixed liquor. In effect, the tracer method is an indirect method for measuring oxygen transfer. A series of grab samples are collected and counted in a scintillation counter.

Gas bubbles must not form in the samples because the krypton-85 would come to equilibrium with the gas bubbles and produce an error in the radioactive counts. Accurate samples counting is essential for good results. The counting efficiency is about 30% for tritium and about 90% for krypton-85. The tracer technique, using radioactive or stable isotopes, has had limited but increasing application.[10,13]

The tracer method is identical for clean water and respiring system conditions because the tracers used, krypton and tritium, are nonreactive substances. For example, for a batch system with Q_r equal to zero, the governing equation is given by:

$$\frac{dT}{dt} = K_L a_f \bar{\tau} \qquad (16)$$

Where:

$\bar{\tau}$ = The ratio of krypton:tritium in the aeration volume; the initial ratio at $t = 0$ is known.

This approach can also be used for flow-through or continuous testing by accounting for the loss of tracers from the aeration volume via the effluent flow.

Other methods. In addition to the above tests, other procedures may also be used to evaluate biological-treatment aeration systems. For the most part, these procedures have not been extensively used and, therefore, cannot be recommended as primary-respiring system field tests. These tests, however, may be applied in specific treatment operations.

Mass Balance: Activated Sludge Systems. The mass balance approach has been proposed to determine oxygen transfer in operational activated sludge systems. Total oxygen balances must be made on the influent, effluent, and waste-activated sludge. The change in total oxygen across the entire activated sludge system equals the oxygen transferred by the aeration system. Total oxygen measurements should be based on COD, with correction for nitrification.

Major problems with the mass balance technique are associated with satisfactory waste-activated sludge volume measurement, obtaining representative waste-activated sludge samples for analysis, and determining an average system DO value. The high SS concentrations in waste-activated sludge make it difficult to obtain valid data because of large errors caused by minor variations in the aeration volume solids inventory. Although it is possible to achieve accurate mass balance measurements on small-laboratory systems, this technique has limited practical value in field-scale aeration equipment evaluation for activated sludge systems. For further discussion on the use of a mass balance technique, the reader is directed to Ball and Campbell[14] for activated-sludge systems and McKeown and others[15–18] for aerated stabilization basin systems.

Dual unsteady-state method. Mueller and Rysinger[19] recently proposed a procedure for evaluation of OTRs under process conditions. The dual unsteady-state analysis uses unsteady-state DO measurements at a high OTR to raise the DO concentration to a value well above that used for process conditions, followed by a lower OTR where unsteady-state DO measurements are again taken. Field oxygen transfer coefficients, K_La_f, for each condition may be calculated. In addition, field-saturation values can be determined for each transfer rate and oxygen-uptake rate. The method requires a substantial spread in K_La_f values from the high OTR condition to the low OTR condition and constant uptake rates during the test period. Results of tests that used this method have been reported.[5,20,21]

FIELD ANALYSIS FOR TEST METHODS

Successful aeration equipment evaluation under process conditions involves careful measurement of several key parameters. A discussion of these measurements and their importance in field evaluations is covered in Chapter 3 and in the following text.

DISSOLVED OXYGEN

An accurate measurement of the DO concentration in an aeration volume is essential to any aeration-equipment evaluation. For a respiring system testing, direct-reading DO probes seem to be the only practical means of measuring DO concentration in MLSS samples. Following proper probe calibration, considerable care and attention are required to ensure continuous reliable results under field conditions.

Aeration testing may be carried out on an entire tank or an isolated-mixing zone in a test volume. Typically, a minimum of three probes should be used in field testing and, depending on the type of aeration system being tested, unit placement in the test tank can be critical. For example, in aeration volumes with a relatively uniform DO concentration (such as well mixed with respect to DO), probes may be located in the tank without regard to aerator-mixing pattern. For aeration volumes, however, that are not well mixed with respect to DO, probes should be strategically placed around the specific-flow pattern of the aeration device. Typical sampling locations for DO measurement for several types of aeration equipment are illustrated in Figure 8.3.

Depending on the specific application, many of the aeration devices shown in Figure 8.3 will not yield a uniform DO concentration in the tank under test conditions. In addition, the flow patterns shown in

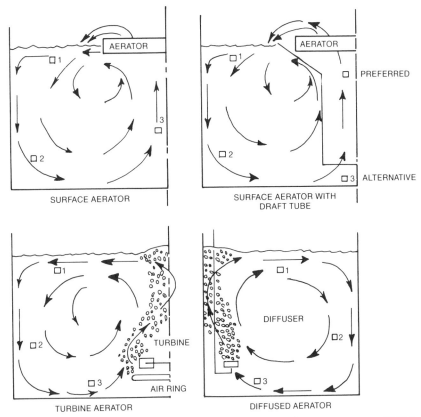

FIGURE 8.3—Typical sampling locations for measurement of DO for different aeration systems.

Figure 8.3 are typical but are not necessarily valid for all applications. Therefore, establishing a complete DO profile on the aeration system before testing is important. The DO profile will indicate actual test-tank mixing characteristics and allow proper DO probe placement. Further discussion on mixing and basin geometry effects on mixing can be found elsewhere.[22]

During aeration-system field testing, DO concentration should not be the limiting factor in the biological reaction although DO can become limiting at approximately 0.5 mg/L for nonnitrifying activated sludge systems and 1.5 mg/L for nitrifying systems. Therefore, it is essential that the minimum DO be above 0.5 mg/L and 1.5 mg/L, respectively, during testing of nonnitrifying and nitrifying activated sludge systems.

OXYGEN UPTAKE RATE

A significant factor for aeration equipment evaluation in respiring systems is an accurate measurement of the MLSS oxygen uptake rate (R). Experience shows that accurate measurement of rapid oxygen uptake created by high-organic loads is virtually impossible. Ideally, oxygen-uptake rate measurements should be taken *in-situ* or immediately at the point of sample collection. As a practical matter, however, a finite-time period elapses before and during field measurements of this parameter. Because the oxygen-uptake rate of a sample will vary as the available soluble substrate is oxidized, significant variations in the uptake

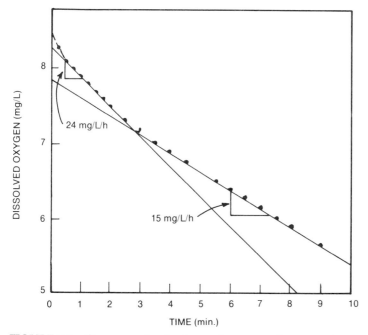

FIGURE 8.4—Oxygen uptake plot for steady state continuous testing, in a moderately loaded activated sludge system, showing a changing uptake rate with time.

rate are likely for samples taken from moderately to highly loaded systems. The information presented in Figure 8.4 illustrates such a situation. Thus, caution is urged where oxygen-uptake rates are being measured under actual plant-loading situations.

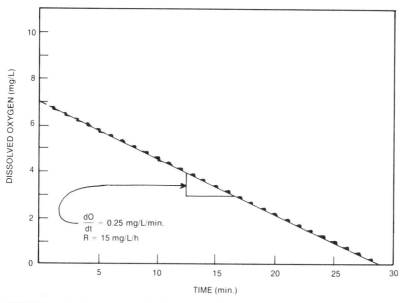

FIGURE 8.5—Oxygen uptake plot for steady state batch endogenous test.

A practical approach used to minimize oxygen-uptake rate variability is to test under endogenous respiration conditions. Endogenous respirations can be achieved by diverting the influent wastewater to other aeration tanks and by allowing the MLSS to assimilate the remaining soluble organics. This procedure is the basis for batch testing that establishes a low, and relatively constant, oxygen-uptake rate that can be more accurately measured. The data presented in Figure 8.5 illustrate the constant uptake rate obtained under endogenous conditions. The reader is cautioned that low substrate concentrations brought about by this testing mode may increase alpha, α, and yield apparently higher OTRs than will be achieved under normal process conditions.

ALPHA AND BETA FACTORS

Alpha and beta values are necessary for describing the influence of mixed liquor characteristics (e.g., dissolved substances and SS) on the aeration equipment transfer capability in clean water. Generally, alpha and beta measurements are only made when the field transfer rates are to be compared to standardized transfer rates developed under clean water conditions.

Because of influent wastewater quality variability, mixed liquor characteristics, and type of aeration device, the alpha level may be extremely variable for a given application. Testing under endogenous respiration conditions is advantageous because it minimizes variability during the test period; however, the OTR obtained is not necessarily the same as that which occurred during operating conditions under normal plant loading. The subject area involving alpha and beta corrections is rather complex. The reader should refer to Chapter 3 and elsewhere for a discussion of the factors affecting alpha and beta and procedures for their determination.[4]

WASTEWATER TEMPERATURE

Wastewater temperature also affects field OTR evaluation. Both OTR and oxygen-uptake rate are significantly altered by temperature change. Biological reactions are thought to be governed by an approximate doubling of rate for every 10°C increase in temperature in the practical operating range of 10 to 30°C.[22] Thus, oxygen-uptake rates should be measured at the actual treatment system-operating temperature. Furthermore, because aeration equipment is rated at 20°C, but is actually tested at some other temperature, appropriate transfer rate adjustments must also be applied to account for wastewater temperature effects. The temperature correction for oxygen transfer is discussed in Chapter 3 and elsewhere in more detail.[4] Where practical, testing at temperatures near 20°C is desirable because it will minimize temperature corrections and will allow a more accurate comparison with transfer rates at standard conditions.

REFERENCES

1. Hwang, H.J. and Stenstrom, M.K., "Evaluation of Fine-Bubble alpha Factors in Near Full-Scale Equipment." *J. Water Pollut. Control Fed.*, 52, 1142 (1985).
2. "A Standard for the Measurement of Oxygen Transfer in Clean Water." Am. Soc. of Civ. Eng., Oxygen Transfer Standards Committee, New York, N.Y. (1984).
3. Barnhart, E.L., "An Overview of Oxygen Transfer Systems." *Proc. Workshop on Aeration System Design, Operation, Testing, and Con-*

trol, U.S. EPA, EPA-600/9-85-005, Cincinnati, Ohio (January 1985).

4. "Development of Standard Procedures for Evaluating Oxygen Transfer Devices." Section 7—Oxygen Transfer Measurements in Respiring Systems, Am .Soc. of Civ. Eng., Oxygen Transfer Standards Committee, EPA-600/2-83-102 U.S. EPA, Cincinnati, Ohio (October 1983).

5. Mueller, J.A., "Nonsteady State Field Testing of Surface and Diffused Aeration Equipment." Manhattan College, Bronx, N.Y. (1983).

6. Downing, A.L. and A.G. Boon, "Oxygen Transfer in the Activated-Sludge Process." In: *Advances In Biological Waste Treatment.* W.W. Eckenfelder, Jr., and B.J. McCabe (Eds.), Pergamon Press, New York, N.Y. (1963).

7. Mueller, J.A., *et al.,* "Oxygen Transfer in Closed Systems." EPA-600-9/78-02, U.S. EPA, Washington, D.C. (1979).

8. Redmon, D., Boyle, W.C., and Ewing, L., "Oxygen Transfer Efficiency Measurements in Mixed Liquor Using Off-Gas Technique." *J. Water Pollut. Control Fed.,* **55,** 1338 (1983).

9. Ewing, L., "New Directions." Proc. Seminar Workshop on Aeration System Design, Testing. Operation and Control, EPA-600/9-85-005.

10. Campbell, Jr., H.H., "Oxygen Transfer Testing Under Process Conditions." Proc. Seminar Workshop on Aeration System Design, Testing, Operation and Control, EPA-600/9-85-005, U.S. EPA, Washington, D.C. (1985).

11. Boyle, W.C. and Campbell, H.J., Jr., "Experiences with Oxygen Transfer Testing of Diffused Air Systems Under Process Condition." Proc. of International Association Water Pollution Control Workshop on Design and Operation of Large Wastewater Treatment Plants, Vienna, Austria (1983).

12. Neal, L.A. "Use of Tracers for Evaluation of Oxygen Transfer." OPA-600-9/78-021, U.S. EPA, Washington, D.C. (1979).

13. Hovis, J.S., and McKeown, J.J., "Gas Transfer Rate Coefficient Measurement by Stable Isotope Krypton/Lithium Technique. I. Treatment Plant Studies." Proc. Intl. Symp. Gas Transfer Water Surface, Cornell U., Ithaca, N.Y.

14. Ball, R.O., and Campbell, H.J., Jr., "Static Aeration Systems—Problems and Performance." Purdue Univ., Lafayette, Ind., *Proc. 29th Ind. Waste Conf. Ext. Ser.,* 328 (1974).

15. Benedict, A.H., and McKeown, J.J., "Oxidation Analysis of Mill Effluents." *Stream Improvement Bulletin NCASI,* **256,** 33 (1972).

16. "A Manual of Practice for Biological Waste Treatment in the Pulp and Paper Industry." *Stream Improvement Bulletin NCASI,* **214,** 115 (1968).

17. McKeown, J.J., Buckley, D.G., and Gellman, I., "A Statistical Documentation of the Performance of Activated Sludge and Aerated Stabilization Basin Systems Operating in the Paper Industry." Purdue Univ., Lafayette, Ind., *Proc. 29th Ind. Waste Conf.,* Purdue Univ., Ext. Ser. 1091 (1984).

18. Alferova, L.A., *et al.,* "Sewage Treatment in the Northern Areas of the U.S.S.R. Report on International Symposium on Wastewater Treatment in Cold Climates." Rep. EPS 3-WP-74-3, 64 (1974).

19. Mueller, J.A., and Rysinger, J.J., "Diffused Aerator Testing Under Process Conditions." Purdue Univ., Lafayette, Ind., *Proc. 36th Ind. Waste Conf.,* Purdue Univ., Ext. Ser., 747 (1981).

20. Mueller, J.A., "Comparison of Dual Nonsteady State and Steady State Testing of Fine Bubble Aerators at Whittier Narrows Plant, Los Angeles," Paper presented at Workshop on Aeration System Design, Operation, Testing, and Control, Madison, Wis. (1982).

21. Mueller, J.A., *et al.* "Dual Nonsteady State Evaluation of Static Aerators Treating Pharmaceutical Wastes." Purdue Univ., Lafayette, Ind., *Proc. 37th Ind. Wase Conf.*, (1982).

22. McKinney, R.E., "Mathematics of Complete Mixing Activated Sludge." *J. Sanit. Eng. Div., Proc. Am. Soc. Civ. Eng.*, **88,** 87 (1962).

Chapter 9
Aeration In
Special Systems

PREAERATION AND POSTAERATION

In addition to providing dissolved oxygen for aerobic biological wastewater treatment processes, aeration is used to control certain wastewater treatment problems through preaeration and postaeration.

Preaeration is performed to achieve one or more of the following objectives:

1. odor control,
2. grease separation and increased grit removal,
3. prevention of septicity,
4. grit separation,
5. flocculation of solids,
6. maintenance of DO in primary treatment tanks at low flows, and
7. minimization of solids deposition on side walls and floors of wet wells.

Both diffused and mechanical aeration devices are used for preaeration. Details on design, types, and performance of various diffused- and mechanical-aeration systems are covered in earlier chapters.

The practice of postaeration[1-3] has evolved in recent years because more stringent water quality standards are being adopted by various regulatory agencies. Although federal secondary-effluent criteria for publicly owned treatment facilities do not include DO, many local discharge requirements specify a minimum DO concentration ranging from 2 to 8 mg/L, depending on the stream requirement. Generally, DO concentration requirements of 2 to 4 mg/L are desirable for secondary effluents, while DO concentrations of 6 to 8 mg/L might be needed for advanced waste treatment systems.

The effluent from most treatment processes does not normally have the DO level that receiving waters require. Consequently, postaeration must be used before the wastewater effluent may be discharged. Although little work has been done in the area of wastewater postaeration, the mechanisms by which treated effluent may be oxygen enriched include the application of reaeration systems using either diffused air, mechanical aeration, spillways or weirs, and waterfalls.

A spillway or waterfall system may be attractive if a significant elevation is available between the treatment end and the receiving water. If land availability is limited, a more compact system such as diffused air or mechanical aeration would usually be more applicable. The choice of a particular reaeration method is site specific.

DIFFUSED AERATION

The design theory of diffused-air systems has been presented in earlier sections of this manual. Reaeration may be accomplished with the use of coarse-, medium-, or fine-bubble systems. Using the same diffuser type at the same depth as is used in the main plant is desirable although remote systems sometimes require a separate local blower and piping system.

The major difference between the design of an effluent reaeration system and an activated sludge aeration system is the inherent inefficiency of oxygen transfer in wastewater effluent. Effluent reaeration systems are normally designed to raise the DO in the effluent to a point close to saturation and under conditions that are not conducive to efficient oxygen transfer, i.e., warm effluent temperatures and low DO saturation. In addition, the oxygen uptake rate of the effluent is low compared to that of the aeration mixed liquor, and is normally ignored in design. This results in a lower DO concentration gradient and lower driving force for oxygen transfer; thus, a low oxygen transfer efficiency. Although the alpha factor of the effluent must be considered in design, it generally will be very similar to that of the clarified aeration tank effluent.

MECHANICAL AERATION

Mechanical aeration for reaeration is particularly appropriate in plants where a central air blower system is not available or where the installation is remote from the main plant operations. Any type of surface aerator may be used and the choice of equipment will dictate the most appropriate tank geometry. In many cases floating, high-speed units are economically preferable.

The theory and basis of mechanical aeration systems and diffused air systems design has been presented earlier in this manual. Details for designing preaeration and postaeration systems with diffused and mechanical aeration equipment are found elsewhere.[1-6]

WEIRS AND SPILLWAYS

In cases where there is a significant difference in elevation between the end of the treatment system and the receiving water, it may be possible to use the available elevation head for reaeration. Such systems are economical because they require no additional power and they are usually easy to maintain.

Free-falling weir systems are more efficient than spillways. The most efficient system is achieved through the creation of turbulence (increased water/air interface) when the water drops from level-to-level without flowing smoothly over the weir face. A spillway system's efficiency may increase if the roughness of the channel increases or if hydraulic jumps occur.

Weir systems. In a weir system, reaeration occurs during water-surface formation at the weir's crest in free fall. The transfer of oxygen is enhanced by entrainment and splash at the lower water surface. Studies by Nakasone[7] have indicated that reaeration over a single weir can be estimated by a series of equations of the form:

$$\ln r_{20} = K(D + 1.5 H_c)^n Q_w^p H_t^r \tag{1}$$

Where:

r_{20} = $(C_{s20}^* - C_o)/(C_{s20}^* - C)$

C_{s20}^* = tabular value of oxygen surface saturation at 20°C, mg/L

C_o = DO concentration of postaeration influent, mg/L

C = DO concentration of postaeration effluent, mg/L

D = drop height from weir edge to water surface, m

H_c = critical water depth on the weir, m

Q_w = wastewater flow rate per unit of weir width, m³/h·m

H_t = tail-water depth, m

K,n,p,r = coefficients depending on Q_w and $(D + 1.5\ H_c)$

This series of equations, which have been tested under laboratory and field conditions, have upgraded the formulations developed by numerous investigators to include tail-water depth, fall height, and discharge rate. Observations indicate that aeration efficiencies are higher for fall heights under 1.2 m (4 ft), thus multiple cascades with less than 1.2 m fall heights are preferable to single falls with greater than 1.2 meters fall heights.

Spillways. If the oxygen demand of the waste is small, the reaeration through a spillway may be defined by

$$D = D_l e^{K_{L}at} \qquad (2)$$

Where:

D = the dissolved oxygen deficit at any point on the spillway, $C_s^* = C$, mg/L,

D_l = the initial dissolved oxygen deficit, $C_s^* - C_o$, mg/L

K_La = the overall oxygen transfer coefficient, $\dfrac{1}{t}$, and

t = the time of travel.

Many formulae have been proposed to estimate K_La for streams. The following equations seem most applicable to spillways and small, high-velocity streams:[8-10]

$$K_La = 10.92\ (V/H)^{0.85}$$
$$K_La = 12.81\ V^{0.5}\ H^{-1.5}$$
$$K_La = 9500\ V\ S$$

Where

V = the average velocity, fps,

H = the average water depth, ft, and

S = the slope of the energy gradient, ft/ft.

The velocity may be calculated from the Manning formula.

$$V = (1.486/n)\ R^{0.67}\ S^{0.5} \qquad (3)$$

Where:

n = 0.018 to 0.022 for a rough concrete channel, and

R = the hydraulic radius, wetted area/wetted perimeter.

Reaeration on a spillway is not very intense. A 60-m (200-ft) length of spillway at 1:1 slope will reduce the deficit of dissolved oxygen from saturation by approximately 25%. Roughening the channel bed or adding obstructions to cause turbulence will approximately double the transfer of oxygen.

*H*IGH PURITY OXYGEN

The concept of using high-purity oxygen in the activated sludge process was first looked at in 1934 by Dr. L.I. Dana, then director of the Linde Research Laboratory. The subject was revived after World War II with the advent of relevantly low-cost, simple oxygen generation plants.

In 1969, the first full-scale demonstration plant was set up at Batavia, New York, as a retrofit in one-half of an existing air-activated sludge plant. This pilot project demonstrated certain process advantages achievable with oxygen aeration as well as with the economical use and dissolution of the oxygen gas.

Since that time, the technology has progressed to a point where the high-purity oxygen activated sludge treatment system is recognized as a significant advancement in the waste treatment field.[11-13] More than 200 plants around the world are now in operation treating both industrial and municipal wastewaters.

The early development and commercial application of the oxygen-activated system was provided by Union Carbide. Firms now providing high purity oxygen technology include the Lotepro Corporation, Inc. (current owners of the Union Carbide system), Air Products and Chemicals, Inc., and Zimpro Inc., purchasers of the system originally developed and marketed by FMC.

High-purity oxygen significantly increases the driving force available for mass transfer from the gas to the mixed liquor; consequently, dissolution rates increase above those normally possible for conventional air systems.

Henry's law states that the saturation concentration of a gas in a liquid is directly proportional to the partial pressure of the gas in the atmosphere in contact with the liquid.

For pure oxygen in water at 0°C, Henry's law constant is 70.5 mg/L per atmosphere. At the normal concentration of oxygen in air of 21% (P equal to 0.21 atmospheres) the equilibrium saturation concentration of oxygen in water at 0°C is 14.8 mg/L.

With high-purity oxygen, the proportion of oxygen increases from 21% to almost 100%, therefore, increasing the equilibrium saturation concentration and raising the saturation concentration by approximately 4.7 times. This is the most significant advantage of using high-purity oxygen in wastewater treatment.

MIXING AND OXYGEN REQUIREMENTS

For the high-purity oxygen activated sludge process to function, it is essential that the following criteria are met:

1. satisfactory DO levels be maintained in the mixed liquor,
2. adequate bulk mixing and liquid-circulation patterns be maintained in the aeration zone to dissolve the oxygen, and
3. mixed-liquor velocity be sufficient to suspend the biological solids.

In terms of activated sludge kinetics, high-purity oxygen systems behave similarly to air-activated sludge systems. The design engineer requires specific knowledge of the chemical and biological constituents in the wastewater, the biomass synthesis rate, and the endogenous-respiration rates to maintain satisfactory DO levels.

Solids suspension velocities of 0.1 to 0.2 m/s (0.3 to 0.5 fps) at tank floor are considered the minimum values for effective suspension of the biological solids that typically range from 3000 to 8000 mg/L. Submerged turbines, mechanical aerators, and submerged-rotating diffusers generally are used to obtain these results.

PROCESS DESCRIPTION

Pure oxygen systems are manufactured and sold as either "closed" or "open." Schematic diagrams of open and closed tank high-purity oxygen systems are shown in Figures 9-1 and 9-2, respectively.

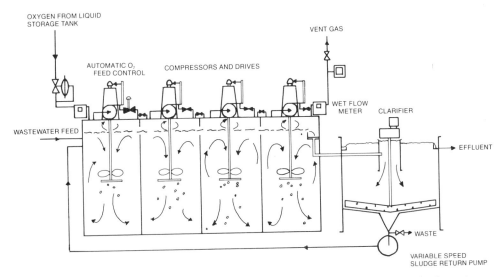

FIGURE 9.1—Closed tank high purity oxygen system schematic.

Closed Tank Systems. With the closed multistage contacting system, the tanks are covered; the oxygen-feed gas is introduced in the first stage and then maintained at a slight positive pressure of approximately 2.5 to 10 cm (1 to 4 in.) of water column. Successive stages are connected by ports at the surface to allow gas to flow from stage-to-stage. Exhaust gases are removed by venting from the last stage of reactor and are monitored for oxygen purity. A relatively simple pressure-regulation system enables the oxygen generator output to be automatically controlled and respond to changes in system oxygen demand as a consequence of the oxygen use by the biomass. As the pressure decreases, the pressure controller opens the valve to allow oxygen to flow into the reactor to increase the pressure to the set level and vice versa.

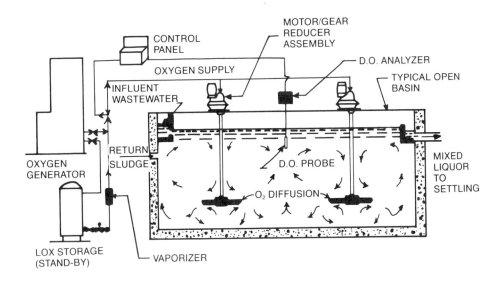

FIGURE 9.2—Open tank high purity oxygen system schematic.

Open Tank Systems. With the open-tank system, oxygen control results from monitoring DO in the mixed liquor and controlling the oxygen supply to the rotating diffuser discs that dissolve the oxygen.

With high-purity oxygen systems, organic loadings and biomass loading ranges of 0.8 to 3.2 kg/m^3·d (50 to 200 lb BOD$_5$/1000 cu ft/day) and 0.3 to 1.5 kg BOD$_5$/kg MLVSS/day, respectively, are possible and depend on effluent quality target and the specific wastewater to be treated.

Because of the high solids concentration carried in the high-purity oxygen system and the high sludge-settling rates, the clarifier hydraulic overflow rate and the solids loading must match the anticipated operating conditions of the reactor. The peak overflow rates typical of final clarifier design range from 20 to 40 m/d (500 to 1000 gal/sq ft/day) and solids loadings from 120 to 240 kg/m^2·d (24 to 50 lb TSS/sq ft/day).

Return sludge rates typically range from 30 to 50%.

OXYGEN GENERATION

The on-site production of oxygen can be provided either by the distillation of air at cryogenic temperatures or through adsorption units.

The cryogenic oxygen generation plants are normally more efficient than the adsorption units. However, adsorption systems are better suited to smaller installations where oxygen use is under 30 tons/day. For purposes of general comparison, an efficiency of approximately 0.033 kW/kg·h (0.15 kW/hr/lb) of oxygen generated is representative of cryogenic plant performance; a figure of approximately 0.44 kW/kg·h (0.20 kW/hr/lb) of oxygen generated is typical of adsorption units.

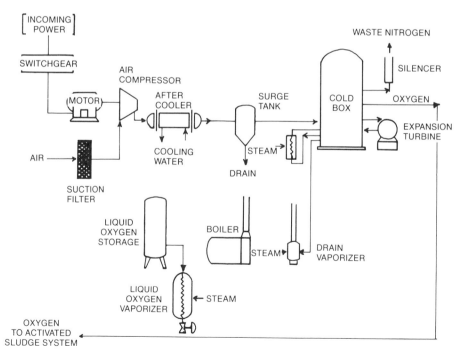

FIGURE 9.3—Cryogenic oxygen system schematic.

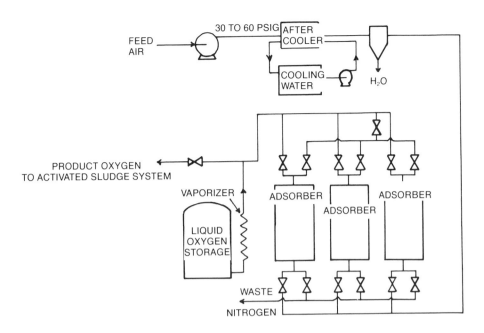

FEED AIR

30 TO 60 PSIG

AFTER COOLER

COOLING WATER

H₂O

PRODUCT OXYGEN TO ACTIVATED SLUDGE SYSTEM

VAPORIZER

LIQUID OXYGEN STORAGE

ADSORBER

ADSORBER

ADSORBER

WASTE

NITROGEN

FIGURE 9.4—Pressure swing adsorption system schematic.

Design of either oxygen generation system for oxygen supply turn down is important. The limitation in adsorption system turndown is the air feed compressor. With proper selection of the air compressor system, turndown to 20% of the full load is possible. With a cryogenic oxygen generation system, turn down is a function of both the air compressors and the distillation-column design; normally it is not less than 70% of the full-load condition. Figures 9.3 and 9.4 show typical cryogenic and adsorption plant-flow diagrams.

Depending on the size and the number of generation units, both systems may require liquid oxygen storage for peak oxygen supply and for downtime of compression equipment.

REFERENCES

1. Kormanic, R.A., "Simplified Mathematical Procedure for Designing Post Aeration Systems." *J. Water Pollut. Control Fed.* **41,** 1956 (1969).
2. Kormanic, R.A., "Design of Plug Flow Post Aeration Basins." *J. Water Pollut. Control Fed.* **42,** 1922 (1970).
3. Lin, S.H., "Axial Dispersion in Post Aeration Basins," *J. Water Pollut. Control Fed.* **51,** 985 (1979).
4. "Wastewater Treatment Plant Design." Manual of Practice No. 8, Water Pollut. Control Fed., Washington, D.C. (1977).
5. Speece, R.E., and Orosco, R., "Design of U-Tube Aeration System." *J. San. Engr. Div.*, Am. Soc. Civ. Eng., **96,** 715 (1970).
6. "Process Design Manual for Upgrading Existing Wastewater Treatment Plants." U.S. EPA, Office of Technology Transfer, Washington, D.C., (1974).

7. Nakasone, H., "Study of Aeration at Weirs and Cascades." *J. San. Engr. Div.*, Am. Soc. Civ. Eng., **113,** 64 (1987).

8. O'Connor, D.J. and Dobbins, W.E., "The Mechanism of Reaeration in Natural Streams." Proc. Am. Soc. Civil Eng., 82 (SA6) 1115-1, (1956).

9. O'Connor, D.J., *et al.*, "Water Quality Analysis of the Delaware River Estuary." *J. San Eng. Div.*, Am. Soc. Civil Eng., Vol. 94 (SA6), 6318 (Dec. 68).

10. Tsivoglou, E.C., *et al.*, "Tracer Measurement of Stream Reaeration, II. Field Studies." *J. Water Pollut. Control Fed.*, 40, 2, 285 (1968).

11. "Activated Sludge: A Comparison of Oxygen and Air Systems." Amer. Soc. Civil Engr., New York, N.Y. (1983).

12. Stenstrom, M.K., *et al.*, "Estimating Oxygen Transfer Capacity of a Full-Scale Pure Oxygen Activated Sludge Plant." Pres. 60th Annual Conf., Water Poll. Control Fed., Washington, D.C. (Oct. 1987).

13. Parker, D.S., and Merrill, M.S., "Oxygen and Air Activated Sludge: Another View." *J. Water Pollut. Control Fed.*, **48,** 2511 (1976).

Symbols and Nomenclature

Symbol	Definition	Unit Dimensions*
ACFM	Actual volume at specified conditions	L^3/t
BHP	Break horsepower of prime mover at the shaft	FL/t
BRV	Bubble-release vacuum	F/L^2
C	Average spatial dissolved oxygen concentration	M/L^3
C^*	Dissolved oxygen saturation concentration	
C^*_T	Dissolved oxygen saturation concentration at a given pressure and temperature, T	
C^*_∞	Spatial average steady state dissolved oxygen saturation concentration approached at infinite aeration time	
$C^*_{\infty T}$	Spatial average dissolved oxygen concentration at a given pressure and temperature, T, approached at infinite aeration time	
C^*_{sT}	Tabular values of dissolved oxygen surface saturation concentration at temperature, T, 1.0 atmosphere and 100% relative humidity	
DWP	Dynamic wet pressure	F/L^2
d_c	Effective saturation depth at infinite time	L
e_b	Blower efficiency	
e_d	Driver efficiency	
e_g	Gear box efficiency	
e_m	Motor efficiency	
f	Used as a subscript to denote process (or field) conditions	
G_i	Mass rate of inerts in air including nitrogen, argon, and trace elements	L^3/t
H	Static head	L
h_L	Diffuser headloss corrected to 20°C	L
K	Transfer rate	
K_La^*	True volumetric mass transfer coefficient in clean water	t^{-1}
K_La	Apparent spatial average volumetric mass transfer coefficient in clean water	
K_La_T	Apparent volumetric mass transfer coefficient at temperature, T	
K_La_f	Apparent volumetric mass transfer coefficient in process water	
M	Molecular weight	L
M_o	Molecular weight of oxygen	L
M_i	Molecular weight of inerts in a gas including nitrogen and argon	L
MR o/i	Mole ratio of oxygen to inerts in feed air	
MR og/i	Mole ratio of oxygen to inerts in off-gas	
N	Field aeration efficiency	$\dfrac{M}{FL}$
N_o	Standard aeration efficiency	$\dfrac{M}{FL}$
n	Rotational rotor speed	t^{-1}
n_s	Specific rotor speed	—
OTE	Oxygen transfer efficiency	
OTE_f	Oxygen transfer efficiency under field conditions	
OTR	Oxygen transfer rate	M/t
OTR_f	Oxygen transfer rate under field conditions	M/t
o	A subscript to indicate a starting time of zero	
P	Power	$F\,L/t$

Symbols and Nomenclature

Symbol	Definition	Unit Dimensions*
p_a	Ambient pressure	F/L^2
p_b	Atmospheric pressure at base conditions	F/L^2
p_s	Standard total pressure of 1 atm. at 100% relative humidity	F/L^2
p_{vt}	Saturated vapor pressure of water temperature, T	F/L^2
Q	Volumetric flow rate	L^3/t
q	Gas flow rate	L^3/t
q_i	Gas flow rate at inlet	
q_o	Gas flow rate at outlet	
R	Rate of oxygen consumption	$M/L^3 t$
RH	Relative humidity	
s	Used as a subscript to denote surface saturation values	
SAE	Standard aeration efficiency	M/t FL/t
SCFM	volume under standard conditions	L^3/t
SOTE	Standard oxygen transfer efficiency	
SOTR	Standard oxygen transfer rate	M/t
T	Temperature	
V	Liquid volume of the aeration system	
W	Oxygen mass transfer rate per unit volume	$M/L^3 t$
W_{O_2}	Mass flow rate of oxygen	M/t
WHP	Wire horsepower of prime mover	FL/t
Y_R	Mole fraction of oxygen in inlet gas	
Y_{OG}	Mole fraction of oxygen in off-gas	
α	Ratio of $K_l a$ in process water to $K_l a$ in clean water at such equivalent conditions as geometry, temperature, and mixing	
β	Ratio of C_∞^* in process water to C_∞^* in clean water at equivalent conditions of temperature and partial pressure	
Ω	Pressure correction factor	
ρ	Density of dry air at standard temperature and pressure	
τ	Temperature correction factor	
θ	Empirical temperature correction factor	
γ_{wT}	Mass density of water at the temperature, T	M/L^3
$\hat{\ }$	A superscript used to indicate the parameters at a particular location with the volume of liquid being aerated	

Standard Condition for Reporting Air Volume, Flow Rate, and Density

1.00 atm
20°C
36% RH

Standard Conditions for Dissolved Oxygen Saturation Concentration

1.00 atm total pressure at surface
20°C
100 RH in gas (mole fraction of oxygen in gas must be specified)

Symbols and Nomenclature

Symbol Definition	Unit Dimensions*

Metric Conversions

Customary Unit   *SI Unit*

psi	×	6.895	=	kN/m^2
psi	×	0.0703	=	kgf/cm^2
sq ft	×	0.0929	=	m^2
sq in.	×	645.2	=	mm^2
tons (short)	×	907.2	=	kg

*Units: M = mass
          L = length
          F = force
          t = time

Table of Dissolved Oxygen Saturation Values

Solubility of oxygen (mg/L) in water exposed to water-saturated air at atmospheric pressure = 101.3 kP$_a$.

Temp	Chlorinity		
°C	0	5.0	10.0
0.0	14.62	13.73	12.89
1.0	14.22	13.36	12.55
2.0	13.83	13.00	12.22
3.0	13.46	12.66	11.91
4.0	13.11	12.34	11.61
5.0	12.77	12.02	11.32
6.0	12.45	11.73	11.05
7.0	12.14	11.44	10.78
8.0	11.84	11.17	10.53
9.0	11.56	10.91	10.29
10.0	11.29	10.66	10.06
11.0	11.03	10.42	9.84
12.0	10.78	10.18	9.62
13.0	10.54	9.96	9.41
14.0	10.31	9.75	9.22
15.0	10.08	9.54	9 03
16.0	9.87	9.34	8.84
17.0	9.67	9.15	8.67
18.0	9.47	8.97	8.50
19.0	9.28	8.79	8.33
20.0	9.09	8.62	8.17
21.0	8.91	8.46	8.02
22.0	8.74	8.30	7.87
23.0	8.58	8.14	7.73
24.0	8.42	7.99	7.59
25.0	8.26	7.85	7.46
26.0	8.11	7.71	7.33
27.0	7.97	7.58	7.20
28.0	7.83	7.44	7.08
29.0	7.69	7.32	6.96
30.0	7.56	7.19	6.85
31.0	7.43	7.07	6.73
32.0	7.31	6.96	6.62
33.0	7.18	6.84	6.52
34.0	7.07	6.73	6.42
35.0	6.95	6.62	6.31
36.0	6.84	6.52	6.22
37.0	6.73	6.42	6.12
38.0	6.62	6.32	6.03
39.0	6.52	6.22	5.93
40.0	6.41	6.12	5.84

NOTE. Benson, B. B., and Krause, D. Jr., "The Concentration and Isotopic Fractionation of Oxygen Dissolved in Fresh Water and Seawater in Equilibrium with the Atmosphere." Limnology and Oceanography (1984).

Index

I

Immediate oxygen demand, 8
Impeller aerators, 84
Inert-gas tracer method, 137
Inlet design, porous diffusers, 55
Inorganic COD, 8
Installation, diffusers, 54
Instrumentation, 105

J

Jet aeration, 30

K

Kraft paper, alpha, 92
Krypton-85, tracer, 143

L

Layouts, non-porous diffusers, 29
Limestone diffusers, 2
Low-speed aerators, 74

M

Maintenance, diffused air systems, 116
Maintenance, diffusers, 50
Maintenance, mechanical aerator systems, 128
Mechanical aeration, 152
Mechanical aeration, 2, 73
Mechanical aerator classification, 75
Mechanical aerators, O&M, 95
Media characteristics, diffusers, 32
Media, diffusers, 22
Metal plate diffusers, 2
Metric conversions, 161
Mixing, 12
Mixing, air requirements, 49
Mixing, diffusers, 46
Mixing, mechanical aerators, 92
Modeling, oxygen transfer, 15
Mounting, diffuser, 52
Municipal wastewater, alpha, 92

N

NBOD, 5
Nitrification, 100
Nitrogenous oxygen demand, 5, 7
Nitrogenous oxygen demand, calculations, 10
Nomenclature, 159
Non steady-state tests, 141
Nonporous diffusers, 27

O

Odor control, 151
Off-gas testing, 137, 142
Omega factor, 18
Open-turbine aerators, 79
Operation, diffused air systems, 115
Operation and maintenance, 111
Operation, mechanical aerator systems, 122
Oxygen generation, 156
OTE, 22, 34, 39, 43, 50, 112
OTR, 15, 89, 112, 134
Outlet design, porous diffusers, 55
Oxygen
  balance, 6
  control, 96
  demand model, 7
  demand, 5, 132
  requirements, 5, 154
  requirements, aeration tank, 10
  rules of thumb, 5
  systems, 153
  transfer, 91
  transfer capacity, surface aerator, 89
  transfer coefficient, 15, 17
  transfer efficiency, 22
  transfer modelining, 15
  transfer rate, 15
  transfer tests, 141
  uptake curves, 12
  uptake rate, 145
O&M, 111
O&M considerations, mechanical aerators, 95

P

Perforated pipe diffuser, 27
Perforated hose diffuser, 29

V

Valved orifice diffuser, 28

W

Water depth, 36
Weirs, 152